ANATOMY AND PHYSIOLOGY
FOR NURSES
including notes on their clinical application

EVELYN C. PEARCE

S.R.N., R.F.N., S.C.M., M.C.S.P., Member R.S.H.

Formerly Senior Nursing Tutor, The Middlesex Hospital, and Member of the General Nursing Council for England and Wales (for many years Examiner in Nursing to the Council, and Examiner in Fever Nursing and Epidemiology for the Diploma in Nursing, London University)

other books by Miss Pearce

A General Textbook of Nursing
Medical and Nursing Dictionary and Encyclopaedia
Instruments, Appliances and Theatre Technique
Nurse and Patient
Environmental Health and Hygiene

ANATOMY AND PHYSIOLOGY FOR NURSES

including notes on their clinical application

by

EVELYN C. PEARCE

with new illustrations by Audrey Besterman

Sixteenth Edition

with 224 illustrations

Wolfe Publishing Ltd

Published by
Mosby-Year Book Europe Ltd
Lynton House
2–17 Tavistock Square
London WC1H 9LB

Reprinted 1992, by Cox & Wyman Ltd, Reading, Berkshire
Reprinted 1993

First published in 1929 by Faber and Faber Limited. Fourteenth
edition 1962, reprinted 1964, with additions 1966. Fifteenth
edition 1968, reprinted 1970, with additions 1973. Sixteenth
edition (reillustrated) 1975, reprinted 1976, 1981, 1983, 1987 and
1989.

ISBN 0 7234 1832 2

For full details of all Mosby titles please write to Mosby-Year Book
Europe Ltd, Lynton House, 2–17 Tavistock Square, London
WC1H 9LB, England.

A CIP Catalogue record for this book is available from the British Library.

PREFACE TO THE FIFTEENTH EDITION

The opportunity to prepare a new edition of this little book on Anatomy and Physiology provided the occasion to have a new look at the contents in order to make it as acceptable as possible to students using the New Syllabuses of the General Nursing Councils.

With this in view, a comprehensive revision includes the Clinical Application of these Basic Sciences—notes are interspersed throughout the text, supplemented by Clinical Notes grouped together at the end of most chapters.

I am indeed fortunate and most grateful for the help given me, in this revision, by competent modern authorities:

E. M. Campbell, Esq., B.Sc., Ph.D., M.D., F.R.C.P., Senior Lecturer to the Royal Postgraduate Medical School and Consultant Physician to Hammersmith Hospital, Editor of *Clinical Medicine*.

Leon C. L. Gonet, Esq., M.B., F.R.C.S., Consultant Orthopaedic Surgeon, Battersea, Putney and Tooting Bec Group of Hospitals.

David I. Hamilton, Esq., M.B., B.S., F.R.C.S., Consultant Cardiac Surgeon to the Liverpool Regional Hospital Board at Broadgreen and Sefton Hospitals.

P. H. F. Silver, Esq., Professor of Embryology, The Middlesex Hospital Medical School, London, W.1.

The new illustrations are by Mrs. Audrey Besterman.

I have pleasure too in thanking the publishers, Messrs. Faber and Faber, for their courteous assistance, and very particularly Miss P. Jean Cunningham, Editor of Nursing and Medical Textbooks, for her expert and generous help in seeing this new edition through the press.

Evelyn C. Pearce

PREFACE TO THE SIXTEENTH EDITION

We are very fortunate to have had the services of Mrs. Audrey Besterman who has entirely reillustrated the new edition of Evelyn Pearce's *Anatomy and Physiology for Nurses*, which has been a standard textbook for nurses for 46 years. The book has always been kept up-to-date, and for this edition Iain Miller, F.R.C.S., has read and amended the text where necessary. For this edition also we have added a glossary of eponyms.

P.J.C.

CONTENTS

LIST OF ILLUSTRATIONS

Chapter 1

INTRODUCTION TO THE HUMAN BODY

Anatomy is the study of the structure of the body and of the relationship of its constituent parts to each other. In *regional anatomy* a geographical study is made and each region, e.g. arm, leg, head, chest, etc., is found to consist of a number of structures common to all regions such as bones, muscles, nerves, blood vessels and so on. From this study it follows that a number of different systems exist. These have been grouped together and described under the heading *systematic anatomy*.

A study of the position and relationship of one part of the body could not be separated from a consideration of the use of each structure and system. This study led to the employment of the terms *functional anatomy* which is closely allied to the study of physiology. Then again it was found that certain structures could be examined by the naked eye and the term *macroscopic anatomy* was introduced to describe this study, in distinction to *microscopic anatomy* which necessitates the use of a microscope. Closely allied to the study of anatomy are *histology*, the study of the fine structures of the body, and *cytology*, the study of the cells.

Physiology is the study of the functions of the normal human body. It is closely linked with the study of all living things in the subject of *biology*; as well as this there is the work of the cytologist, interested in details of the structure of cells, and that of the *biochemist*, dealing with the chemical changes and activities of cells and investigating the complex chemistry of life, and there is *physics*, the study of the physical reactions and movements taking place in the body.

The body is made up of many tissues and organs, each having its own particular function to perform. The *cell* is the *unit* or the smallest element of the body of which all parts are comprised. The cells are adapted to perform the special functions of the organ or tissue they are in. Some cells, such as those in the nervous system and muscle, are very specialized indeed; others, such as those in the connective tissues, are less highly developed. As a general rule the most highly specialized cells are the least able to withstand damage and also are the most difficult to repair or replace.

Terms used in Anatomy. Many parts of the body are *symmetrically* arranged. For example the right and left limbs are similar; there are right and left eyes and ears, right and left lungs, and right and left kidneys. But there is also a good deal of *asymmetry* in the arrangement of the body. The spleen lies entirely on the left side; the larger part of the liver lies on the right side; the pancreas lies partly on each side.

The human body is studied from the erect position with the arms by the sides and the palms of the hands facing forwards, the head erect and eyes looking straight in front. This is described as the *anatomical position*.

The various parts of the body are then described in relation to certain *imaginary lines* or *planes. The median plane* runs through the centre of the body. Any structure which lies nearer to the median plane of the body than another is said to be *medial* to that other. For example the hamstring muscles which lie on the inner side of the thigh are nearer the median plane than those which lie on the outer side and are therefore *medial* to the other group which are described as *lateral*. Similarly the inner side of the thigh is described as the *medial aspect* and the outer as the *lateral aspect*.

The terms *internal* and *external* are used to describe the relative distance of an organ or structure from the centre of a cavity. The ribs for example have an internal surface which is near the chest cavity and an external surface which is on the outer side, farther away from the cavity. The internal carotid artery (*see* Fig. 11/4, page 181), is within the cranial cavity and the external is outside the cavity.

The terms *superficial* and *deep* are used to denote relative distance from the surface of the body, and the terms *superior* and *inferior* denote positions relatively high or low, particularly in relation to the trunk, such as the superior and inferior surfaces of the clavicle.

The terms *anterior* and *posterior* are synonymous with *ventral* and *dorsal*. These terms are only applied to man in the erect attitude or 'anatomical position'. For example the anterior and posterior tibial arteries lie in front and behind in the leg.

In describing the hand the terms *palmar* and *dorsal* are used instead of anterior and posterior, and in describing the foot the terms *plantar* and *dorsal* are similarly employed.

The terms *proximal* or *distal* are employed to describe nearness to, or distance from a given point, particularly in relation to the limbs. For example the proximal phalanges are nearer to the wrist and the distal ones are the farthest away. When three structures are in a line running from the median plane of the body outwards, they are described as being placed in *medial*, *intermediate* and *lateral* positions. An example of this is seen in the arrangement of the three cuneiform bones of the foot (*see* page 109). Similarly when three structures run from front to back (anterior to posterior) or from above downwards (superior to inferior), these are described as anterior, middle and posterior as happens in the arrangement of the three fossae of the skull (*see* Fig. 3/1, page 64); and superior, middle and inferior, as occurs in the arrangement of the superior, middle and inferior radio-ulnar joints (*see* Fig. 7/7, page 118).

THE SYSTEMS OF THE BODY

Systematic anatomy or the division of the body into systems is arranged (*a*) according to the functions they perform and (*b*) under the heading of the different terms employed to indicate the knowledge of certain parts.

Osteology	is a knowledge of		bones,
Arthrology	,, ,,	,,	,, joints,
Myology	,, ,,	,,	,, muscles,
Splanchnology	,, ,,	,,	,, organs or viscera,
Neurology	,, ,,	,,	,, nerves and nerve structure

When grouped according to function the general arrangement is as follows.

The Locomotor System. This includes the parts concerned in the movements of the body: the *skeletal system* which is composed of the bones, and certain cartilages and membranes, the *articulatory system* which deals with the joints or articulations, and the *muscular system* which includes muscles, fascia and tendon sheaths (*see* Chapters 3 to 8).

The Blood-Vascular System includes the *circulatory system* and *lymphatic system*. Blood is the principal transport system; it is pumped round the body by the heart, oxygen is brought from the lungs and carbon dioxide collected from the tissues. Food passes to the liver and thence to the general circulation. Waste products are passed to the kidneys.

The Digestive System consists of the alimentary canal and the glands and organs associated with it. Food is broken down by enzymes in the digestive tract and taken by the blood to the liver and finally to the tissues.

The Respiratory System contains the passages and organs concerned with breathing. Oxygen from the air is taken into the blood and carried to the tissues; the waste product, carbon dioxide (CO_2), is carried by the blood from the body tissues to the lungs and breathed out in the expired air.

The Ductless Glands are grouped together because of the internal secretions they produce. The spleen is sometimes included in this group because it also has no duct, though as far as is known it does not produce an internal secretion; it is concerned with the formation of red blood cells and is described in Chapter 10.

The Urogenital System includes the organs of the *urinary system* and the *reproductive system*. The waste products of the body, except carbon dioxide, are excreted by the kidneys.

The Nervous System is composed of the *central nervous system* which includes the brain and spinal cord, the *peripheral nervous* *system* consisting of the nerves given off from brain and cord and the *autonomic nervous system*. The central and peripheral systems are often grouped together and described as the *cerebrospinal nervous system*. The autonomic nervous system includes the *sympathetic* and *parasympathetic nerves*. It is also described as the *involuntary nervous system*.

The Special Sense Organs include taste, smell, sight and hearing, and also the tactile function of the skin. It is through these organs that the individual is kept aware of external forces and thus enabled to protect himself. A chicken aware of the sound of traffic runs or flies to safety.

The Excretory System is the term sometimes employed to describe collectively the organs that deal with the excretion of waste products from the body. These organs include the *urinary system* (*see* above), the *lungs* in their function of eliminating carbon dioxide, and the *colon* which excretes certain insoluble substances in the faeces.

THE BODY FLUIDS

Water with its solvents needed for the health of the cells is termed body fluid, and this fluid is partly inside and partly outside the cells.

Intracellular fluid forms 50 per cent of the body-weight; it lies within the cells, and contains electrolytes including potassium and phosphates and food materials like glucose and amino-acids. Enzyme action is constant within the cells, breaking down and building up as in all metabolism to maintain a balance.

Extracellular or interstitial fluid represents 30 per cent of the water in the body (about 12 litres). It is the medium in which the cells live, obtaining from it salts, food, and oxygen and passing into it their waste products.

Blood plasma forms 5 per cent of body-weight (about 3 litres) and it is the transport system which serves the cells through the medium of the extracellular fluid.

Tissue Fluid Exchange. The fluid of the plasma is under greater mechanical *hydrostatic pressure* than the interstitial pressure and therefore fluid tends to leave the capillaries. However, there are proteins in the plasma but not in the interstitial fluid; these plasma proteins exert an *osmotic pressure* which tends to suck fluid into the capillaries.

At the arterial end of the capillaries the mechanical, hydrostatic pressure is greater than the osmotic pressure so the balance of the forces sends fluids out into the tissues. At the venous end the hydrostatic pressure is less; the osmotic pressure overcomes it and draws fluid back into the capillaries. Normally there is more fluid leaving the capillaries than there is fluid coming back into them. This excess is removed by the lymphatics.

Exchange between the extracellular and intracellular fluids is also dependent on osmotic pressure, but the cell membrane too has a selective permeability, allowing some substances, such as oxygen, CO_2 and urea to cross freely, but pumping others either in or out to maintain different concentrations in the intra- and extracellular fluids. For example, potassium is concentrated in the intracellular fluid, whilst sodium is pumped out.

Clinical Notes

Fluid and electrolytic balance. Normally the same amount of fluid is taken into the body as is excreted from it. Water and electrolytes are taken in as water, as other liquids and in food. Water is normally lost from the kidneys, as urine; it is lost from the skin in sweat, from the alimentary tract in the faeces and from the lungs by the saturated air breathed out. Electrolytes are lost in urine, from the skin and from the alimentary tract. The adaptability of the body to maintain electrolytic balance is phenomenal. There is increased kidney activity to respond to an increased fluid intake, and the warning thirst of a man who has lost excess water in sweating, indicates his need for fluid.

The management of a patient's fluid and electrolyte balance is of great importance, as either depletion or excess can have serious consequences.

Dehydration or depletion of the body fluids is of two types: Lack of water as in shipwrecked sailors. It causes thirst, mental disturbance and fever. Dehydration also occurs in infants, helpless people, particularly the elderly, and in unconscious patients who are not given enough water. Lack of salt (sodium) is more important. It is caused by excessive loss from the body as in diarrhoea and vomiting. It causes shrinkage of the tissues, faintness, loss of blood pressure and muscular weakness. It does not cause thirst.

In *shock* the pulse is rapid, the skin clammy, the volume of circulating blood is decreased and the blood pressure is low. The commonest causes are haemorrhage and salt depletion.

Sodium depletion occurs in excessive sweating, and cannot be corrected by drinking water alone. When uncorrected it leads to muscle cramps, loss of energy, fatigue and faintness. It is seen in people entering a hot climate from a cooler one, and in those working under conditions of great heat. It can be corrected by having dilute saline drinks or taking some salt tablets until acclimatized.

Sodium excess occurs in renal failure, and when too much saline is given intravenously.

Potassium is another essential electrolyte. *Depletion* occurs in many medical conditions, as in protracted vomiting, in loss of fluid from an ileostomy and during treatment by diuretics unless these are supplemented by potassium given by mouth.

Water intoxication may occur in patients given too much water without sodium, as, for example, glucose and water, and who are unable to excrete it. The blood sodium concentration is reduced (which may be mistaken for sodium depletion), and the patient becomes confused and has convulsions.

Fluid balance chart. The above notes outline the basic principles underlying the urgency of accurately recording the fluid intake and output when requested, for upon it not only the well-being of a patient, but even his life, may depend. This necessitates a factual consideration of all fluid intake in liquid and food (*see* page 205), and in any fluid given artificially, as well as an exact consideration of all fluid output, including discharges, bleedings, gastric and bronchial suctions, in vomiting and diarrhoea, in colostomy and ileostomy effusions, from a wound, superficially as in burns, from other injuries and from surgery.

OEDEMA

Oedema is waterlogging of the tissues due to a breakdown of the delicate balance described above. It can obviously arise from one of four reasons:
  (1) Too high a mechanical hydrostatic pressure in the capillaries as happens, for example, if the venous drainage is blocked.
  (2) Too low an osmotic pressure due to insufficient plasma protein, particularly albumin.
  (3) Blockage of the lymphatics.
  (4) Damage to the capillary walls so that the plasma proteins leak out into the tissue and cause an osmotic pressure opposing the osmotic pressure of the protein in the blood stream.

Cardiac oedema occurs in congestive heart failure (*see* page 160). The venous pressure and consequently the capillary pressure is raised. Oedema in the legs and feet occurs in those who habitually stand and walk, over the sacrum in those who sit, and the lower part of the back and buttocks in those who lie. The kidneys are involved and the secretion of sodium is diminished. *An important factor causing oedema* is the inability of the kidney to excrete sodium.

Oedema due to Lymphatic Obstruction is most characteristically, though not invariably, seen in the arm after radical mastectomy, as the surgeon has removed the lymph glands which drain the axillary area (*see* page 196). This form of oedema also occurs in *elephantiasis* due to *filariasis* caused by a tropical parasite which blocks the lymphatics.

Oedema is also seen in *thrombosis of the deep veins of the leg* which is a dangerous complication of prolonged confinement to bed and which allows the blood flow to become too sluggish so that clots form. It may also be the result of infection.

THE TISSUE CELL

A cell is a minute (jelly-like) mass of protoplasm containing a nucleus held together by a cell membrane. In considering the *structure of a cell* its component parts may usefully be related to its function (*see* page 23).

Cells possess the qualities of all living matter, including those of self-preservation and reproduction.

Ingestion and assimilation. Cells select from the *intercellular* or *interstitial fluid which surrounds them* chemical substances such as *amino-acids*, which the cell builds up into the very complicated substances, e.g., proteins, which make up protoplasm. Thus a cell is a very active unit in which the nourishing food materials eaten by man are absorbed and assimilated.

Growth and repair. These materials brought to the cell may be used by the cell to synthesize new protoplasm, in which case the cell increases in size, that is, it *grows*. They may also be used to replace worn-out parts of the cell. These constructive activities, growth and cell repair, are spoken of as the anabolic functions of the cell, or *anabolism*.

Metabolism. On the other hand, the cell needs a supply of *energy* for its activities and it will use some absorbed food materials as a fuel. The food is broken down (*catabolism*) and the energy stored in it is released and used by the cell to provide heat, glandular secretion, movement, and nervous activity. Anabolism and catabolism make up the total activities of the cell, or *metabolism* (*see also* Chapter 17).

Respiration. Oxygen brought from the lungs by the blood and the removal of the gaseous waste-product, carbon dioxide, are essential for the functions and survival of the cell.

Excretion. The *waste materials* resulting from the catabolic processes are eliminated from the cell into the interstitial fluid, and thence carried away by the blood. The blood transports the carbonic acid waste to the lungs where it is removed from the body as carbon dioxide. The other waste substances are eliminated, via the kidneys, in the urine.

Irritability and Conductivity. Mention has been made of some of these cells' characteristic functional properties, their metabolic activities and powers of growth.

By these two properties the cell is active. When a cell is stimulated either by chemical, physical, mechanical, or nervous means, the cell responds; it may contract as does a muscle cell (fibre); it may produce a secretion, as do the cells of the stomach, pancreas, and other organs and glands; or it may conduct an impulse, as in the case of the nerve cell. This last is the best example of cell conductivity as a nerve impulse generated by the stimulation of a nerve cell may be conducted for a considerable distance, a yard or more, according to the length of the nerve fibre. But in all cases, a stimulus which *excites* a cell to action is *conducted* along the entire length, from end to end of the cell.

Cell Structure. In considering the structure of the cell it is essential to relate its parts to its **function**.

The *protoplasm of the cell* is composed of a centrally placed body, the *nucleus*, and the *cytoplasm* or remainder of the protoplasm, which surrounds the nucleus.

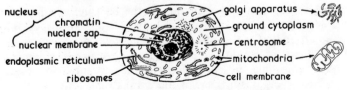

FIG. 1/1.—THE DIFFERENT PARTS OF A TISSUE CELL

Cytoplasm. This contains the following essential requirements:

1. *Mitochondria*, small rod-like structures which are closely connected with the catabolic, or respiratory, processes of the cell body.

2. *Golgi apparatus*. A canal-like structure lying next to the nucleus and involved in the secretory activities of the cell.

3. *Ground cytoplasm*. A highly complex colloidal material in which the other structures are embedded. It is largely concerned with the anabolic, or synthetic, activities of the cell.

4. *Centrosome*. A minute dense part of the cytoplasm, lying close to the nucleus. It plays an important part during cell division (*see* Fig. 1/2).

5. *Cell membrane*. The cell boundary is no static envelope. Many important functions are connected with it, but in particular it acts as a selective sieve through which certain substances are allowed to pass into the cell, or which prevents other substances from gaining access to it. Thus it is most important in maintaining the correct chemical composition of protoplasm.

Nucleus. The nucleus consists of a more compact mass of protoplasm, separated from the cytoplasm by the *nuclear membrane* which is also selectively porous, allowing substances to escape from the nucleus into the cytoplasm or substances to pass into it. The nucleus controls the cell and all its activities. Without a nucleus the cell would die.

The nucleus contains many protein-rich threads lying in a nuclear sap. In the 'resting cell' the threads are collectively spoken of as *chromatin*. These threads or *chromosomes* are vital to the everyday activities of the cell and are responsible for determining the hereditary characteristics of the human body. On the chromosomes in linear arrangement sit the genetic or hereditary determinants, the *genes*. The number of chromosomes in a body cell is constant for a particular species of organism. In man there are twenty-three pairs of forty-six chromosomes (*see further*, p. 321).

Reproduction. A cell does not go on growing indefinitely in size but at a certain optimum point divides into two daughter cells. Further, certain cells will undergo division to replace worn-out cells or those destroyed by disease. This kind of cell division is called *mitosis*, or *karyokinesis*.

Activity begins in the nucleus, the nuclear membrane disappears and the *chromatin* changes character and becomes long filaments called *chromosomes*. The *centrosome* divides and the *two new centrosomes* move away from each other to each end of the nucleus called the poles. The chromosomes are then

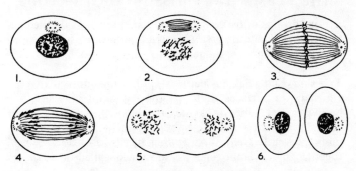

FIG. 1/2.—STAGES IN CELL DIVISION

1. Cell with Nucleus and Centrosome.
2. Nucleus changes. Centrosome divides.
3. Spindle Fibres in position.
4. Two identical sets of Chromosomes being attracted to the Poles.
5. and 6. Two new Cells separating.

attracted to the poles and lie near the new centrosomes. The *chromatin* of which the nucleus is formed now comes to rest and *two new nuclei* exist. Finally the *protoplasm of the cell constricts and divides* and the two new cells are complete (*see* Fig. 1/2).

Each new daughter cell resulting from mitosis contains forty-six chromosomes, so that during mitosis each chromosome must duplicate itself. The process of chromosomal duplication is one of the least understood of the cell's activities.

However, mitosis is not the only kind of cell division. In the sex organs, the ovary and testis, another kind of cell division occurs called *meiosis*. During the formation of the sex cells, or gametes, the number of chromosomes is halved, so that the spermatozoon contains only twenty-three chromosomes and the egg-cell, or ovum, twenty-three.

When fertilization occurs, that is when spermatozoon and ovum fuse to form the cell (zygote) which develops into a new individual, the normal chromosomal complement of forty-six is restored. By this means a mixing of the hereditary determinants, or genes, from male and female is achieved (*see also* page 321).

THE ELEMENTARY TISSUES OF THE BODY

Four groups of tissue in the body are known as the elementary tissues. These are *epithelial* tissue, *muscular* tissue, *nervous* tissue, and *connective* tissue.

Types of Epithelial Tissue. An epithelium consists of cells which cover surfaces of the body, e.g. skin; or which line hollow organs, tubes or cavities, e.g. blood vessels, and the air cells. There are two main classes of epithelial tissue, each containing several varieties. All epithelial cells lie on and are held together by a homogeneous substance called a *basement membrane*.

Simple Epithelium. This class consists of a single layer of cells, and is subdivided into three varieties.

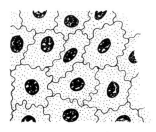

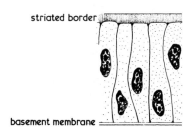

FIG. 1/3.—PAVEMENT CELLS OF SQUAMOUS EPITHELIUM

FIG. 1/4.—COLUMNAR EPITHELIAL CELLS FROM INTESTINE

Pavement or Squamous Epithelium. Pavement epithelial cells are fine thin plates placed edge to edge like the particles in a mosaic pattern or the stones of a pavement. These cells form the alveoli of the lungs. They are found whenever a very smooth surface is essential as in the lining of the heart (serous membrane), lining of blood vessels and lymphatics. When lining these structures the epithelial covering or lining is called *endothelium*.

Columnar Epithelium forms a single layer of cells which line the ducts of most glands, the gall-bladder, nearly the whole of the digestive tract, in which goblet cells are interspersed, and parts of the genito-urinary tract.

The illustration Fig. 1/4 shows columnar cells from the intestine; these have a slightly striated border. In some situations, as when lining the alveoli of secreting glands, the cells of columnar epithelium are short and have a cubical appearance —they are then described as *cubical* cells (*see* Fig. 1/8).

Ciliated Epithelium is found lining the air-passages and their ramifications such as the frontal and maxillary sinuses. It also lines the uterine tubes or oviducts and part of the uterus and the ventricles of the brain.

Ciliated cells are like columnar cells in shape, but they have in addition fine hair-like processes attached to their free edge. These processes are called *cilia*. The ciliary processes keep up a continual movement directed towards the external opening. This movement has been likened to the movement seen in a field of corn, blown in one direction by the wind. In the respiratory passages the constant movement prevents dust, mucus, etc., entering the lungs, and in the uterine tubes the movement conveys the ovum into the uterus.

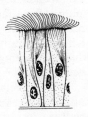

FIG. 1/5.—CILIATED COLUMNAR EPITHELIAL CELLS

Showing hair-like projections.

Goblet Cells are mucus-secreting cells which lie in the walls of glands and ducts lined by columnar cells, either plain or ciliated. Goblet cells secrete mucus or *mucin* and express it on to the surface; they act as mucus-secreting glands and are most numerous where a considerable amount of mucus covers the surface as in the stomach, colon, and trachea.

Compound Epithelium consists of more then one layer of cells. *Stratified Epithelium* forms the epidermal layers of the skin.

It also lines the mouth, pharynx, oesophagus, the lower part of the urethra, the anal canal and the vagina, and covers the surface of the cornea. In these areas it does not become cornified.

The outer layers of cells near the surface comprise the *horny layer* of the skin; these cells are flattened and resemble scales. The deepest layer of cells are columnar in shape. These form the *germinative layer* and here the cells multiply by karyo-kinesis, pushing those above them nearer the surface until the superficial ones are cast off.

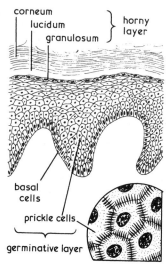

FIG. 1/6.—MICROSCOPIC APPEARANCE OF THE EPIDERMIS—A COMPOUND (STRATIFIED) EPITHELIUM

The cells between the basal layer and the horny zone are called 'prickle cells'; they are connected to each other by fine tendrils which give them a prickly appearance when examined under the microscope (*see also* Chapter 19).

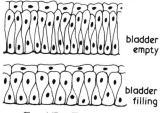

FIG. 1/7.—TRANSITIONAL EPITHELIUM

Transitional Epithelium is a compound stratified epithelium consisting of three layers of cells. It lines the urinary bladder, the pelvis of the kidney, the ureters and the upper part of the urethra. The deeper layers of cells in transitional epithelium are of the columnar type of cell with rounded ends which make them pyriform or pear-shaped. As the cells in the deeper layers multiply by dividing, the superficial layers of cells are cast off.

The superficial cell layers in transitional epithelium are less scale-like than those of stratified epithelium. Comparison of the illustrations Figs. 1/6 and 1/7 will make this point clear.

Functions of Epithelial Tissue. The epithelial tissue which forms the covering of the body, the skin, and the lining of the cavities which open on to the surface is mainly *protective*. It prevents injury to the underlying tissues, prevents the loss of fluid from these tissues and also prevents the passage of fluid into the structures which are covered by skin. Micro-organisms cannot pass through healthy skin but they can and do pass through abraded skin.

Secretory. Most of the secreting glands and their ducts are composed of columnar epithelium. Very often the epithelium

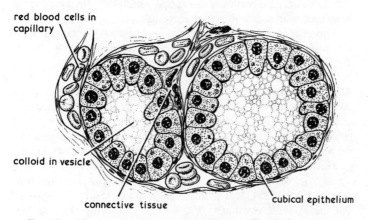

FIG. 1/8.—MICROSCOPIC APPEARANCE OF THYROID GLAND STRUCTURE
The Vesicles are lined with cubical (columnar) Epithelial Cells.

lining the gland and its duct is continuous with that of the surface in which the glands lie. Simple tubular and simple saccular glands are just involutions from the surface such as the simple tubular glands of the intestine as shown in Fig. 1/9. When these involutions branch, the structure becomes more complicated, as in the formation of *compound tubular glands* such as those of the kidney, and *compound racemose* or *saccular glands* such as the salivary glands and the pancreas.

The *endocrine glands* are also composed of epithelial cells which may be massed together or may line hollow vesicles as occurs in the thyroid gland where the vesicles are lined by columnar epithelial cells, cubical in shape. These cells produce their secretion—*colloid*—but there is no duct from these glands and therefore the secretion reaches the blood stream either directly or through the lymphatics.

Glands. A gland is a *secretory organ* which may exist as a separate organ such as the liver, pancreas, and spleen; or may be simply a layer of cells as the *simple tubular glands* of the alimentary canal, body cavities, etc. (*see* Fig. 1/9). All glands have a rich blood supply. Their special function is to select from the blood stream certain substances, which they then elaborate into their important juices or secretions.

There is a tremendous variety of glands, each with its different function, making a collective description and classification difficult. A simple classification is as follows:

Glands which pour their secretion directly on to the surface include the sweat glands, sebaceous glands, and the gastric and intestinal glands.

Glands which pour their secretion indirectly, by means of ducts, on to the surface include the salivary glands, pancreas, and liver.

Ductless glands. These form the group described as endocrine organs (*see* Chapter 18). These are the glands of internal secretion. A great deal of the well-being of the body depends on these glands, which through their secretions exert an important chemical control on the functions of the body.

Membranes. Layers of specialized cells which line the cavities of the body are described as membrane. The three principal membranes are:

Mucous membrane
Synovial membrane
Serous membrane

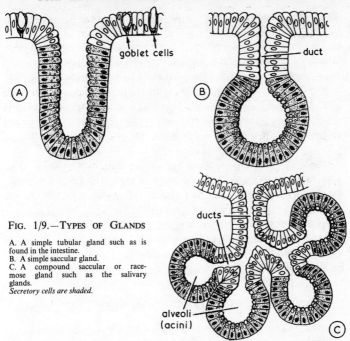

FIG. 1/9.—TYPES OF GLANDS

A. A simple tubular gland such as is found in the intestine.
B. A simple saccular gland.
C. A compound saccular or racemose gland such as the salivary glands.
Secretory cells are shaded.

All these membranes secrete a fluid to lubricate or moisten the cavity they line.

Mucous Membrane is found lining the alimentary tract, the respiratory tract, and parts of the genito-urinary tract. It varies in character in the different areas. In the digestive tract it consists of columnar epithelial cells closely packed together. Some of them become distended with mucous secretion and are then called *goblet cells*. The cell becomes more and more distended and finally ruptures and discharges its secretion on to the surface (*see* Fig. 1/9).

Mucus is the secretion of the membrane and consists of water, salts, and a protein, *mucin*, which gives the sticky or viscid character to the secretion.

Synovial Membrane lines the cavities of joints. It consists of fine connective tissue, with a layer of squamous endothelial cells on the surface. The secretion of synovial membrane is thick and glairy in character.

Serous Membranes are found in the chest and abdomen, covering the organs contained therein and lining these cavities.

The pleura covers the lungs and lines the thorax.

The pericardium covers the heart as a double layer.

The peritoneum covers the abdominal organs and lines the abdomen. (These membranes are described in the chapters dealing with these various organs.)

The characteristics which are common to all three serous membranes are, that each consists of a double layer of membrane having an intervening potential cavity which receives the fluid secreted by the membrane. This *serous fluid* is very similar to blood serum or lymph. It acts as a lubricant, and in addition it contains protective substances and removes harmful products, passing these on to the lymphatic system to be dealt with.

MUSCULAR—NERVOUS—CONNECTIVE TISSUE

Muscular Tissue. Muscle is a tissue which is specialized for contraction, and by means of this, movements are performed. It is composed of cylindrical fibres which correspond to the cells of other tissues. These are bound together into little bundles of fibres by a form of connective tissue which contains a highly specialized contractile element.

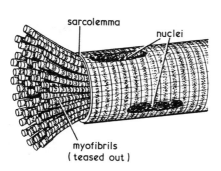

FIG. 1/10.—MICROSCOPIC VIEW OF A STRIPED MUSCLE FIBRE

There are three types of muscle:

Striped (striated, skeletal or voluntary muscle). The individual muscle fibres are transversely striated by alternate light and dark markings. Each fibre is made up of a number of myofibrils and enclosed in a fine membrane—the *sarcolemma* (meaning—muscle sheath). A number of fibres are massed together to form bundles; many of these bundles are bound together by connective tissue to form large and small muscles. When a muscle contracts it shortens, and each individual fibre takes part in the movement by contracting. This type only contracts when stimulated to do so by the nervous system.

Unstriped (unstriated, smooth, or involuntary muscle). This type will contract without nervous stimulation although in most parts of the body its activity is under the control of the autonomic (involuntary) nervous system. This variety is composed of elongated spindle-shaped muscle cells which retain the appearance of a cell (*see* Fig. 1/11).

Involuntary muscle is found in the coats of blood and lymphatic vessels, in the walls of the digestive tract and the hollow viscera, trachea, and bronchi, in the iris and ciliary muscle of the eye, and in the involuntary muscles in the skin.

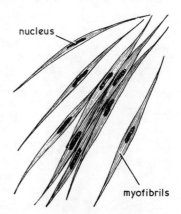

FIG. 1/11.—MICROSCOPIC VIEW OF UNSTRIPED MUSCLE FIBRES

A *sphincter muscle* is composed of a circular band of muscle fibres situated at the internal or external openings of a canal, or at the mouth of an orifice, tightly closing it when contracted. Examples include the cardiac and pyloric sphincters at the openings of the stomach, the ileo-colic sphincter or valve, the internal and external sphincters of the anus and urethra.

Cardiac muscle is found only in the muscle of the heart. It is striated like voluntary muscle. But it differs in that its fibres branch and anastomose with each other; they are arranged longitudinally as in striated muscle, are characteristically red in colour and not controlled by the will.

Cardiac muscle possesses the special property of *automatic rhythmical contraction* independent of its nerve supply. This function is described as *myogenic* as distinct from neurogenic. Normally the action of the heart is controlled by its nerve supply (*see* page 153).

Muscular Contraction. When a muscle is stimulated a short *latent period* follows, during which it is taking up the stimulus. It then *contracts*, when it becomes short and thick, and finally it *relaxes and elongates*.

In the case of a *striped* (voluntary) *muscle fibre* the contraction lasts only a fraction of a second and each contraction occurs in response to a single nerve impulse. Each single contraction is of the same force. The force with which a *whole muscle* contracts is adjusted by varying the number of fibres contracting and the frequency with which each fibre contracts. When contracting vigorously the individual fibres may contract more than 50 times each second.

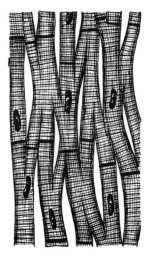

FIG. 1/12.—A MICROSCOPIC VIEW OF CARDIAC MUSCLE SHOWING THE CHARACTERISTIC BRANCHING OF ITS FIBRES

Certain factors influence the force with which a muscle fibre contracts. It contracts more forcibly when it is stretched and when it is warm. Fatigue and cold weaken the power to contract.

Unstriped muscle fibres contract much more slowly and are not dependent on nervous impulses, although these alter the force of their contraction.

Muscle tone. Muscle is never completely at rest; it may appear to be, but it is always in a condition of muscle tone, which means ready to respond to stimuli. For instance, the *knee-jerk* obtained by sharply tapping the patellar tendon results in contraction of the quadriceps extensor of the thigh and slight extension of the knee joint. This is a reflex produced by stimulation of the nerves.

Posture is determined by the degree of muscle tone.

The energy of the muscular contraction is provided by the conversion of adenosine triphosphate (ATP) into adenosine diphosphate (ADP). ADP is then immediately turned back into ATP by energy provided by the breakdown of glycogen. In the presence of adequate supplies of oxygen, this breakdown is aerobic and produces carbon dioxide and water. If there is not enough oxygen, the glycogen is only broken down to lactic acid (anaerobic glycogen) and the content of lactic acid in the blood increases. This is a normal occurrence in vigorous athletes, but occurs too readily in patients whose heart or circulation does not supply the working muscles with enough blood.

Nervous Tissue. The nervous tissue consists of three kinds of matter, (*a*) *grey matter*, forming the nerve cells, (*b*) *white matter*, the nerve fibres, and (*c*) *neuroglia*, a special kind of supporting cell, found only in the nervous system, which holds together and supports nerve cells and fibres. Each nerve cell with its processes is called a *neurone*.

Nerve cells are composed of highly specialized granular protoplasm, with large nuclei and cell walls as other cells. Various processes arise from the nerve cells; these processes carry the nerve impulses to and from the nerve cells. (For further details of the nervous tissue, *see* Chapters 22 and 23.)

Connective Tissues. Connective tissue provides the framework of the body. There are several varieties of connective tissue.

Areolar tissue. This consists of loosely woven tissue which is distributed widely throughout the body. It is placed immediately beneath the skin and mucous surfaces forming the subcutaneous and sub-mucous tissue, and it also forms the sheaths of fascia which support and bind and connect together muscles, nerves, blood vessels, and other organs.

Areolar tissue consists of a matrix of intercellular substance in which lie connective tissue cells and into which are woven bundles of *fine white fibres*, composed of wavy strands, running through the matrix in every direction and so arranged that they form a network. These fibres consist of *collagen*, a gelatinous substance, and they are held together by mucin.

Elastic fibres which are yellow in appearance and composed of elastin also form part of the structure. These fibres are fine.

The tissue spaces in which lymph collects are large, and it is from the lymph contained in them that most of the nourishment of areolar tissue is derived. These lymph spaces communicate with each other, and it is here, in these, that many of the immunizing substances which protect the body from disease are formed.

Retiform (reticular) lymphoid or adenoid tissue is similar to areolar, but a particular kind of cell, the lymphocyte, is present in very large numbers and forms the bulk of the tissue. The lymphocytes are held together by fine connective tissue fibres, called reticular fibres. These are like immature collagen fibres.

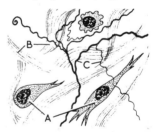

Mucoid tissue is found in the umbilical cord, at birth, in the jelly of Wharton. It is also found in the adult in the vitreous humour of the eye.

FIG. 1/13.—AREOLAR TISSUE

A. Connective Tissue Cells.
B. White Fibres.
C. Elastic Fibres.

Adipose tissue. Adipose or fatty tissue is deposited in most parts of the body. It is associated with areolar tissue by the deposition of fat cells specially adapted for storing droplets of fat, and is present in all subcutaneous tissue except that of the eyelids and the penis, and inside the cranial cavity.

Functions. To help support and retain in position the organs of the body. The kidneys, for example, are deeply embedded in fat.

To form a protective covering for the body.

To act as a store of water and of fat which when required can be re-absorbed and, by combustion in the tissues during metabolism, provide a source of heat and energy for the use of the body.

Elastic tissue. This form of connective tissue contains a large proportion of elastic fibres. It is found in the walls of arteries and in the air tubes of the respiratory tract and assists in keeping these vessels and passages open. It is also present in certain ligaments, as in the ligamentum sub-flava of the vertebral column, where because of its elastic and extensible qualities it materially assists in the performance of sustained muscular effort, as in maintaining the erect position of the spine.

Fibrous tissue is often spoken of as a *white fibrous tissue* because it is composed mainly of white collagen fibres arranged in definite lines. This arrangement gives great strength and fibrous tissue is found where resistance is required. Between the definite bundles of white fibres some areolar tissue lies, which contains the nerves, lymphatics, and blood vessels supplying the structure.

Fibrous tissue is tough and strong. It forms ligaments, except the elastic ones, and tendons. The dura mater lining the skull and neural canal, the periosteum covering bone, the strongest layers of fascia separating muscular sheaths, the fibrous layer of the pericardium, and the sclerotic coat of the eye, are examples of fibrous tissue.

Cartilage or gristle is a dense, clear blue-white substance, very firm but less firm than bone. It is found principally at joints and between bones. The bones of the embryo are first cartilage, then the growing centres persist as cartilage and when adult age is reached cartilage is found covering the bone ends. Cartilage does not contain blood vessels but is covered by a membrane, the *perichondrium*, from which it derives its blood supply.

There are three main varieties of cartilage which demonstrate the characteristics of this substance—firmness, flexibility and rigidity.

Hyaline cartilage consists of collagen fibres embedded in a clear, glassy, tough ground substance or matrix. It is firm and elastic and is found covering the ends of the long bones as articular cartilage, in the costal cartilages, in the nose, larynx, trachea, and bronchial tubes where it keeps open the orifices. It is also the temporary cartilage from which bone is formed. In the developing embryo and fetus it acts as a temporary scaffolding supporting the other tissues until bone is laid down to replace it. The cells of hyaline cartilage are arranged principally in small groups (*see* Fig. 1/14A), set in a tough matrix.

White fibro-cartilage which is composed of bundles of fibres having the cartilage cells arranged between the bundles of fibres is found where great strength is required. Fibro-cartilage deepens the cavities of bony sockets, as in the acetabulum of the innominate bone, and the glenoid cavity of the scapula. It also forms the inter-articular cartilages, as in the semilunar cartilages of the knee, and the connecting cartilages, as in the intervertebral discs of the vertebral column and in the pad of cartilage at the symphysis pubis (*see* Fig. 1/14B).

Elastic cartilage is often called *yellow elastic cartilage* because it contains a great many elastic fibres which are yellow. It is found in the lining of the ear, the epiglottis and the Eustachian (pharyngotympanic) tubes; when compressed or bent it is very flexible and readily springs back into shape (*see* Fig. 1/14C).

Bone Structure and Growth. Bone is the hardest of the connective tissues of the body. It is composed of nearly 50 per cent water; the remaining solid parts are divided into a composition of mineral matter, principally calcium salts 67 per cent, and cellular matter 33 per cent.

The structure of bone may be examined by the naked eye when the *gross structure* is seen, and with the aid of a microscope, when the *minute structure* is examined.

Bone consists of two kinds of tissue: compact tissue and cancellous tissue.

Compact bone tissue is hard and dense; it is found in flat bones and in the shafts of the long bones, and as a thin covering over all bones.

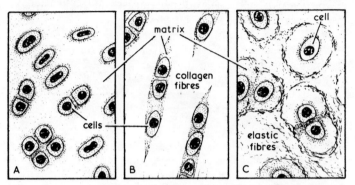

FIG. 1/14.—A. HYALINE CARTILAGE. B. FIBRO-CARTILAGE. C. ELASTIC CARTILAGE

Cancellous bone tissue is spongy in structure. It is found principally in the ends of the long bones, in the short bones, and as a layer in between two layers of compact tissue in the flat bones such as the scapula, cranium, sternum and ribs.

The Gross Structure of a Long Bone (*see* Fig. 1/15). A long bone, such as those of the limbs, shows both varieties of bone tissue. When sawn through longitudinally, the distribution of compact and cancellous tissue can be seen. It is divided into a shaft, the central part, and two extremities or ends of the bone. If the shaft is cut across, dense bone tissue will be seen and a hollow centre called the medullary cavity, containing yellow bone marrow. If the end of a long bone is cut, the spaces in the cancellous tissue will be seen containing red bone marrow. In the yellow marrow fat cells predominate; in the red marrow red blood cells are very numerous The red bone marrow is the birthplace of both red and white blood cells.

Minute Structure. A transverse section of compact bone shows a wonderful design mapped out in circles. In the centre of each circle is a *Haversian canal*. The plates of bone or *lamellae* are arranged concentrically around the central canal, in between these plates are minute spaces called *lacunae*, these spaces contain bone cells and the spaces are connected to

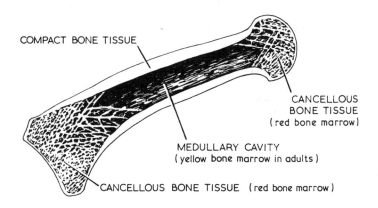

FIG. 1/15.—A LONG BONE CUT LONGITUDINALLY

each other and to the central Haversian canal by minute canals called *canaliculi*. Each pattern so formed is a *complete Haversian system* composed of:

> A central *Haversian canal* containing nerves, blood vessels, and lymphatics,
> *Lamellae* arranged concentrically,
> *Lacunae* containing bone cells, and
> *Canaliculi* radiating between and linking up the lacunae and the Haversian canals.

The areas between these Haversian systems are composed of *interstitial lamellae*, and *canaliculi* arranged somewhat differently. Fig. 1/16 shows how the Haversian canals run longitudinally through the bone and also shows the difference between the Haversian system and inter-Haversian bone structure.

In cancellous bone tissue the lamellae are rather irregularly arranged, there are no Haversian canals, the blood vessels ramify in the interstitial spaces which are filled with marrow to support the minute blood vessels.

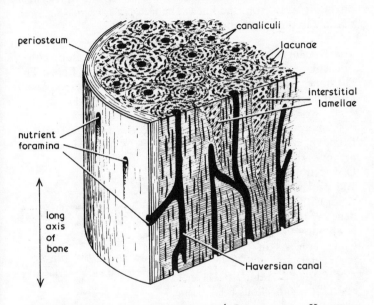

FIG. 1/16.—MICROSCOPIC VIEW OF COMPACT BONE SHOWING HAVERSIAN SYSTEM OF LAMELLAE AND CANALS, ALSO THE INTERSTITIAL ARRANGEMENT OF BONE STRUCTURE BETWEEN THE HAVERSIAN SYSTEMS

Bone is covered by a vascular membrane, the periosteum, but at the articular surfaces it is covered by hyaline cartilage.

Periosteum is a fibrous vascular membrane covering bone. It is rich in blood vessels, and invests the bone closely. The blood vessels from the periosteum ramify in the bone substance, and in this way help to supply the bone with blood. In growing bone a layer of bone-forming cells lies between the periosteum and the bone, and from multiplication of these cells the bone grows in circumference.

In addition to blood derived from the periosteum the long bones are supplied with blood by a special *nutrient artery*, which pierces the bone obliquely at a protected part of it—in the case of the long bones of the upper extremity, in a direction towards the elbow, and those of the lower extremity, in a direction away from the knee. The *nutrient foramina* are well marked in the long bones.

The Development and Growth of Bone. Bone develops either in cartilage or in a membrane of connective tissue fibres. The flat bones ossify in membrane and are called *membrane bones* and the long bones in cartilage, called *cartilage bones*.

Membranous ossification. The connective tissue membrane from which the flat bones develop, such as those of the skull, is very richly supplied with blood. Ossification begins at defined centres, and proceeds by the multiplication of cells within the membrane until a delicate network of bone is formed. Eventually a flat bone is produced which consists of two layers of hard compact bone covered by periosteum, separated by an interstitial layer of bone resembling cancellous bone tissue.

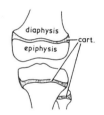

Fig. 1/17.—The Position of the Epiphyseal Cartilage in the Lower Extremity of the Femur and the Upper Extremities of the Tibia and Fibula

Cartilaginous ossification. In the developing embryo all the long bones are first represented by rods of cartilage covered by *perichondrium* (the membrane covering cartilage). A *primary centre of ossification* called the *diaphysis* appears at the middle of what will eventually be the shaft of a long bone. Calcium is laid down in the matrix, and *bone cells* develop. The perichondrium becomes *periosteum* and from it bone cells are laid down so that the bone increases in circumference as well as in length. It is because of this function of the periosteum that surgeons are careful, when removing bone, to replace the periosteum in position, for the formation of new bone from it. The growing bone now consists of a shaft, the *diaphysis* and two extremities, the *epiphyses*.

Later in the process of development, a secondary centre of ossification appears at each extremity or *epiphysis* and ossification begins there and extends towards the shaft and also towards the end of each epiphysis. The ends of the bone remain covered by hyaline cartilage which becomes the articular cartilage. A layer of cartilage remains between the shaft or diaphysis and each extremity or epiphysis, and this layer is called the *epiphyseal cartilage* (*see* Fig. 1/17) which persists until the bone is fully grown.

If *acromegaly* occurs, which is brought about by disorder of the function of the anterior lobe of the pituitary gland (*see* page 283), before the epiphyseal cartilages have disappeared the condition of *gigantism* results; but when acromegaly occurs after the ossification of the epiphyseal cartilage then only certain bones are affected—those of the hands and jaw.

Two types of bone cells are involved in the building of bone; *osteoblasts* which build bone, and *osteoclasts* which destroy bone. In this way the solid parts are formed and the spaces, cavities and canals also are constructed.

Clinical Notes

The development of bone requires a well-balanced diet containing all essential food factors (*see* Ch. 13). Calcium and phosphorus are particularly necessary. An adult requires 1 gram of calcium a day; in pregnancy more is needed as the mother's blood serum must supply the calcium required for bone and tooth formation in the developing fetus. Calcium is obtained from milk, cheese, cabbage, carrots and other vegetables, and phosphorus from milk, egg-yolk and green vegetables (*see* page 206). Foods containing vitamin D which promote the absorption of calcium are essential for bone calcification. Deficiency of vitamin D in the diet of children gives rise to *rickets* because there is insufficient absorption of calcium, resulting in interference with bone calcification and the bones soften. In adults deficiency gives rise to *osteomalacia*.

It is estimated that over 90 per cent of the calcium in the body is contained in the bones and teeth.

Even when fully grown, bone is not an inert passive substance. The cells and chemical constituents are being continually replaced under the influence of hormones and the stresses of weight-bearing and activity. If a patient is kept completely at rest for a long time some of the constituents are mobilized into the blood stream, with consequent weakening of the bone structure.

In *osteoporosis*, as this condition is called, the whole skeleton, particularly the spinal column, is affected, resulting in shortening of the spine and *kyphosis* (rounded back). Osteoporosis may also occur in bones around joints which have been immobilized in plaster casts for long periods.

In *osteitis deformans*, Paget's disease of bone, one or more bones may be affected with a tendency to pathological fracture.

In certain conditions imbalance of the calcium content of bone may cause the bone to soften and bend or, alternatively, to become dense and marble-like. As a rule the balance between calcium taken into the body and the level of its content in bone is maintained by the parathyroid glands (*see* page 279).

Separation of epiphysis. The joint between the bone shaft and bone end may become separated in childhood as a result of traumatic injury. This constitutes the condition of a *slipped epiphysis. Periostitis* is inflammation of the periosteum and this may be associated with infection of the bone tissue, *osteomyelitis*.

Malignant disease. Bone may be, though comparatively rarely is, the site of a tumour, a sarcoma; but it is a common site for deposits of carcinoma.

Chapter 2

SHORT NOTES ON SURFACE ANATOMY

Surface anatomy means the study of 'living' anatomy. All students of anatomy have a rich source of knowledge at their disposal, namely—their own body. This should be used frequently to confirm points learnt from reading and from specimens. With knowledge thus gained the observation and examination of surgeon or physician may be followed with great interest as, by inspection, palpation, percussion and auscultation he studies the condition of his patients in order to assess their needs and arrive at a diagnosis.

The various bony points serve as landmarks; and the position of many of the organs and internal structures is described in relation to these points.

The superficial aspects of the trunk and limbs shown in Figs. 2/5 to 2/13 are a guide to the normal levels of shoulders and hips, the development of muscles and the position of some of the bony prominences which may be sites of soreness in those who lie in bed (*see* shaded areas in Fig. 2/7).

Some Points of Surface Anatomy of the Head. A line drawn from the external occipital protuberance of the occipital bone, forwards over the top of the skull, to a point at the centre of the base of the nose, marks the position of the *longitudinal fissure* separating the cerebral hemispheres, and also the position of the *superior sagittal sinus* of the dura mater (*see* Fig. 11/12, page 190).

The central sulcus or *fissure of Rolando* is marked by finding the mid-point between the external occipital protuberance and the nasion (base of the nose). A line joining this point, but half an inch behind it, and the ear gives the direction of the central sulcus (*see* Fig. 22/7, page 335).

The mastoid process can be felt behind the ear.

The parotid gland lies wedged between the mastoid process and the ramus of the mandible, but it overflows across the masseter muscle beneath the zygomatic arch. Its ducts run forward to pierce the muscle of the cheek (buccinator), to enter the mouth opposite the second upper molar tooth (*see* Fig. 14/5, page 219).

The facial artery passes over the body of the mandible, anterior to the angle (*see* Fig. 11/4, page 181).

The temporal artery crosses the zygomatic process of the temporal bone, in front of the ear.

The Neck. The region of the neck is divided into two main *triangles, anterior* and *posterior,* by the *sternomastoid* muscle

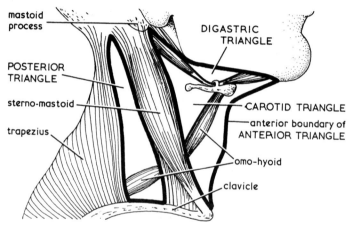

FIG. 2/1.—THE POSITION OF THE TRIANGLES OF THE NECK

which, running obliquely from the mastoid process of the temporal bone to the front of the clavicle, is palpable in its entire length. The *clavicle* lies at the base of the neck, separating it from the thorax.

The *posterior triangle* of the neck is bounded in front by the sternomastoid muscle and behind by the anterior border of trapezius; it contains portions of the cervical and brachial plexuses of nerves, a chain of lymphatic glands which lie posterior to the sternomastoid (*see* Fig. 8/3, page 131), and

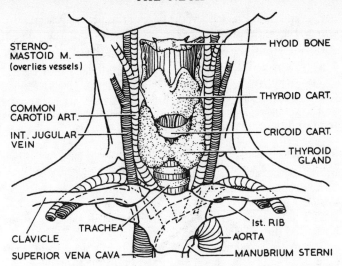

FIG. 2/2.—RELATIVE POSITION OF THE STRUCTURES IN THE NECK

The relative positions of the arch of the aorta and of the superior vena cava are shown as they lie behind the manubrium of the sternum. The common carotid arteries and the jugular veins embrace the trachea. The levels of the hyoid bone, the thyroid gland, the thyroid and cricoid cartilages are indicated.

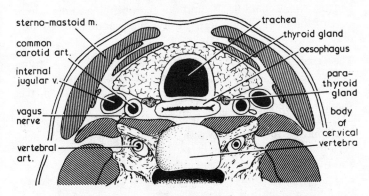

FIG. 2/3.—DIAGRAM OF CROSS-SECTION OF FRONT OF THE NECK

The trachea lies in front embraced by the lobes of the thyroid gland; the isthmus of the thyroid lies in front of the trachea which can be compressed by it when the thyroid gland is enlarged. Blood vessels lie at each side. The oesophagus lies behind the trachea, between it and the vertebral bodies.

nerves and blood vessels. At the base of this lies the first rib over which the subclavian artery passes. It is here that digital pressure can be applied to the subclavian artery.

The *anterior triangle of the neck* is subdivided into several triangles, two of these are shown in Fig. 2/1; the *carotid triangle* is so named because it contains the carotid artery and its divisions into internal and external carotids; it also contains the internal jugular vein, numerous other veins, arteries and nerves. The *digastric triangle* lies below the jaw; it contains parts of the submandibular and parotid salivary glands, a branch of the facial nerve and facial artery, and other structures more deeply placed, including some of the carotid vessels.

The neck from the front. See Fig. 2/2. The *manubrium sterni* is a valuable landmark as behind it lie part of the arch of the aorta and the innominate veins.

The *trachea* commences immediately below the cricoid cartilage (*see* Fig. 2/2, page 47) and passes into the thoracic cavity to terminate by dividing into right and left bronchus at the level of the sternal angle (angle of Louis) (*see* Fig. 2/6, page 53).

The *oesophagus* also begins at the lower border of the cricoid cartilage and runs downward behind the trachea (*see* Fig. 16/3, page 257).

The *thymus* gland lies behind the manubrium and the upper part of the body of the sternum in a child, and in some cases may extend upwards into the neck.

The Trunk. Many of the organs have been described in relation to their surface anatomy, in the description of the position of the organs in the various sections of the book to which the student is referred.

In relation to the vertebral column, *the top of the sternum* lies opposite the joint between the second and third thoracic vertebrae; the *angle of Louis* (*see* page 53) between the fourth and fifth; and the articulation between the body of the sternum and the xiphisternum lies about the level of the disc between the ninth and tenth thoracic vertebrae.

The Anterior Aspect of the Trunk (*see* Figs. 2/5 and 2/6). The *sternal angle* or *angle of Louis* can be felt through the skin; it is the level of attachment of the second rib. At the other end of the sternum is the *infrasternal angle* or *xiphoid* where a shallow depression can be seen and felt. By running the finger from this angle outwards along the lower borders of the ribs, the *costal margin* which is formed by the cartilages of the seventh, eighth, ninth and tenth ribs can be felt. This margin can be seen in thin subjects.

The *apex beat of the heart* can be felt, and sometimes can be seen, in the fifth left intercostal space, 9 cm (3½ inches) from mid-line (*see* Fig. 9/1, page 150 for the position of the heart in relation to the chest wall).

The abdomen is conveniently divided by four planes or imaginary lines, two vertical and two horizontal, into *nine regions* (*see* Fig. 2/4, page 50), so that organs are variously described as lying in part or parts of one or more regions.

The liver, for example, occupies parts of the right hypochondriac and epigastric, extending transversely into the left hypochondriac and occupying also part of the lumbar region.

The *apex of the lung* rises above the level of the clavicle as indicated in Fig. 2/6, page 53 (*see also* Fig. 16/4, page 258) where the position of the lungs, and of the pleura in relation to the heart and the chest wall, will be found indicated.

The space occupied by the lungs at the back is shown in Fig. 2/8, the distance to which the pleura descends is shown stippled. The *spleen* lies on the left side beneath the ninth, tenth and eleventh ribs; the *left kidney* lies between the eleventh thoracic to the third lumbar vertebra, the *right kidney* is slightly lower as its upper pole lies against the liver.

The Abdomen. The *linea alba* forms a depression running down the middle line of the abdomen from the xiphoid cartilage to the symphysis pubis. On each side of this line the rectus abdominis can be felt. Throw this muscle into contraction by lying on your back and with the arms by the sides, elevate the shoulders, and/or raise the lower limbs off the surface on which you are lying. With one hand the recti muscles will easily be felt contracting. The *umbilicus* lies on a level with the disc between the third and fourth lumbar vertebrae.

The *anterior superior spine* (*see* Fig. 2/5) can be distinctly felt, and by drawing a line from the umbilicus to the superior iliac spine on the right side and marking a point at the junction of the middle and outer thirds of this line, McBurney's point, the maximum site of pain in appendicitis, is indicated.

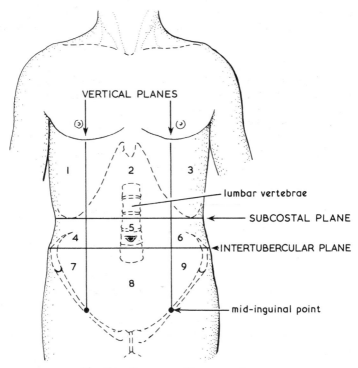

FIG. 2/4.—PLANES AND REGIONS OF ABDOMEN

Regions:
1. The Right Hypochondriac.
2. The Epigastric.
3. The Left Hypochondriac.
4. The Right Lumbar.
5. The Umbilical.
6. The Left Lumbar.
7. The Right Iliac.
8. The Hypogastric.
9. The Left Iliac.

The *stomach* lies in the upper and left aspect of the abdomen partly behind the lower ribs and cartilages; the cardiac orifice lies behind the seventh left costal cartilage. The fundus of the stomach reaches as high as the fifth left intercostal space.

The *position of the liver*, *pancreas*, *duodenum*, *gall-bladder* and *parts of the colon* are indicated in Fig. 2/6. The *gall-bladder* projects slightly from the costal margin at the level of the ninth right costal cartilage; the *pancreas* lies at the back of the abdominal cavity across the first lumbar vertebra; the *aorta* divides into the common iliac arteries in front of the fourth lumbar vertebra; the *caecum* on the right side, and the commencement of the *sigmoid flexure of the colon* on the left lie respectively in the right and left iliac fossae.

The Posterior Aspect of the Trunk (*see* Figs. 2/7 and 2/8). Looked at from the back the *vertebral spines* can be palpated; the *spine of the seventh cervical vertebra* is prominent, *vertebra prominens*, the spine and lower angle of the *scapula* can be felt and seen in thin subjects. The scapula lies in relation to the vertebral column opposite the distance between the spines of the second to the seventh dorsal vertebrae.

The position of the *posterior superior iliac spine* (on each side) can be distinguished by a dimple. The *crest of the ilium* can be palpated in its entire length and at its highest point it lies on a level with the interval between the third and fourth lumbar vertebrae. By marking this line with a skin pencil the position below which it is safe to perform lumbar puncture is indicated. The spinal cord ends at the level of the second lumbar vertebra.

The Extremities (*see* Figs. 2/10 to 2/17 and 22/21 and 22/23 inclusive). Many of the points are dealt with in the various chapters. The different bony points can be felt on palpation.

In the *upper extremity*, the axilla and cubital fossa are described on pages 143–5.

In the *lower extremity*, for Scarpa's triangle and the popliteal space, *see* pages 145–6.

For main arteries, *see* Chapter 11, p. 177.

For principal veins, *see* Chapter 11, p. 187.

For lymphatic vessels, *see* Chapter 12, p. 195.

For principal peripheral nerves, *see* Chapter 22, pp. 352 and 354–6.

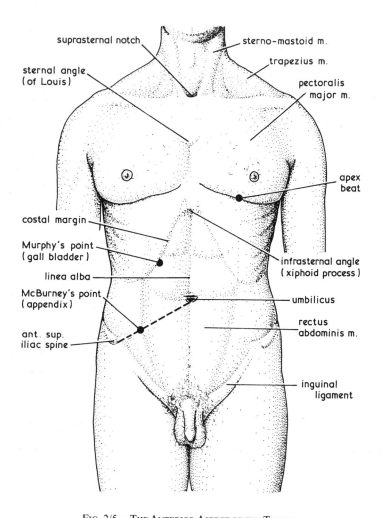

FIG. 2/5.—THE ANTERIOR ASPECT OF THE TRUNK

The suprasternal notch, the sternal angle (of Louis), the xiphoid and the anterior superior spines can be felt.

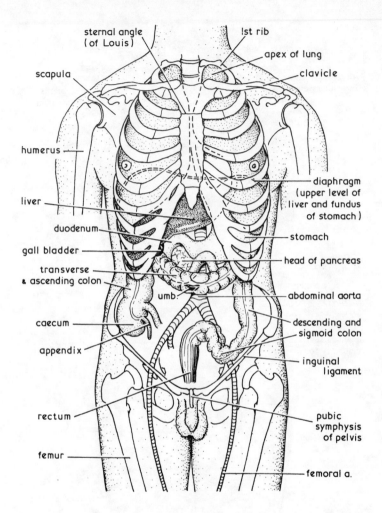

FIG. 2/6.—THE TRUNK, SHOWING THE POSITION OF ORGANS IN RELATION
TO THE ABDOMINAL WALL

For the position of the heart and lungs in relation to the thorax, *see* Fig. 9/1, page 150,
and Fig. 16/4, page 258.

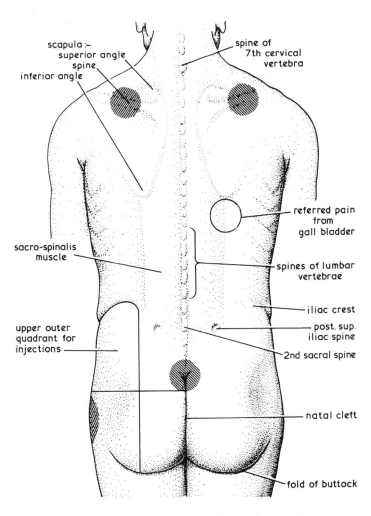

scapula :–
superior angle
spine
inferior angle

spine of
7th cervical
vertebra

referred pain
from
gall bladder

sacro-spinalis
muscle

spines of lumbar
vertebrae

iliac crest

post. sup.
iliac spine

upper outer
quadrant for
injections

2nd sacral spine

natal cleft

fold of buttock

FIG. 2/7.—THE POSTERIOR ASPECT OF THE TRUNK, SURFACE ANATOMY
Areas prone to pressure sores are crosshatched.

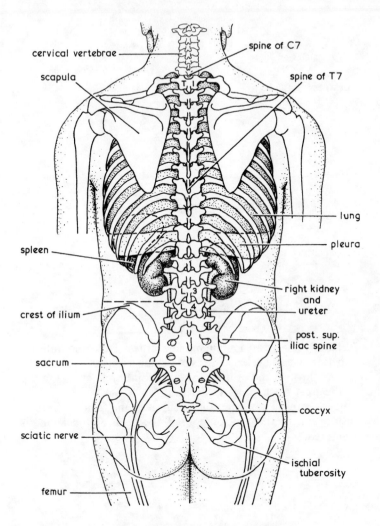

FIG. 2/8.—THE TRUNK FROM THE BACK

The apex of the lung rises above the clavicle; the position of the pleura is shown stippled.

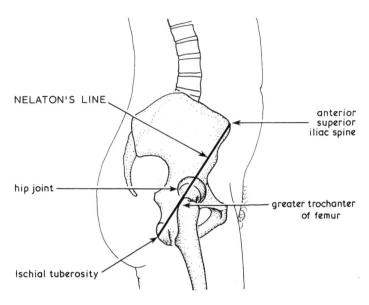

FIG. 2/9.—NELATON'S LINE

Nelaton's line is an imaginary line drawn from the anterior superior iliac spine, backwards to the tuberosity of the ischium. It cuts through the centre of the hip joint and across the top of the great trochanter of the femur.

It is useful in assessing the position of the femoral head in dislocation of the hip joint, or if the neck of the femur is fractured.

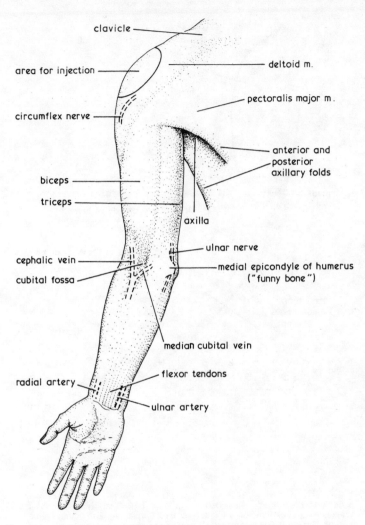

FIG. 2/10.—RIGHT ARM IN ANATOMICAL POSITION

Note 'carrying angle' at elbow joint.

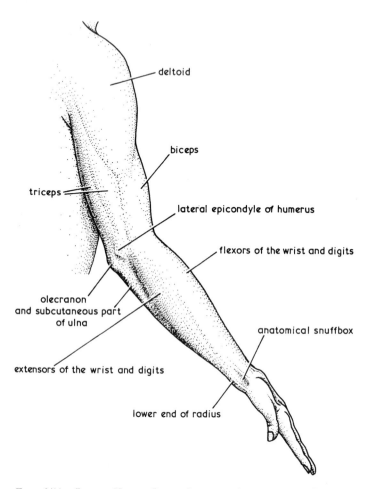

deltoid

biceps

triceps

lateral epicondyle of humerus

flexors of the wrist and digits

olecranon
and subcutaneous part
of ulna

anatomical snuffbox

extensors of the wrist and digits

lower end of radius

Fig. 2/11.—Right Upper Limb, Lateral Aspect, with Forearm
Supinated, Showing the Position of the Principal Muscles of the
Upper Arm. The Thumb is Extended to Render the Extensor Tendons
of the Thumb Taut indicating the Dimple known as the Anatomical
Snuffbox which lies between them

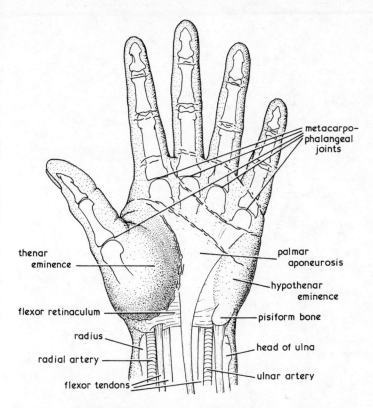

metacarpo-
phalangeal
joints

thenar
eminence

palmar
aponeurosis

hypothenar
eminence

flexor retinaculum

pisiform bone

radius

head of ulna

radial artery

ulnar artery

flexor tendons

FIG. 2/12.—PALMAR ASPECT OF THE LEFT HAND INDICATING THE POSITION
OF THE METACARPO-PHALANGEAL JOINTS, THE THENAR AND HYPOTHENAR
MUSCLE EMINENCES, THE PALMAR FASCIA AND TENDONS CROSSING THE
WRIST BENEATH THE FLEXOR RETINACULUM

It is an important clinical point when applying a splint or plaster to keep the wrist extended, that it should
not impinge on the metacarpo-phalangeal joints but terminate below them, to ensure that the patient can
flex his fingers over the splint, to right angles with the palm of his hand.

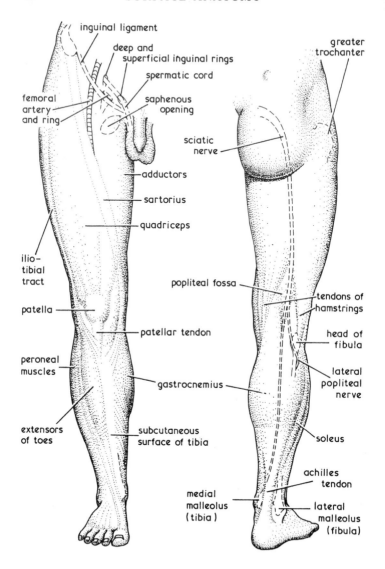

inguinal ligament

deep and
superficial inguinal rings

spermatic cord

femoral
artery
and ring

saphenous
opening

greater
trochanter

sciatic
nerve

adductors

sartorius

quadriceps

ilio-
tibial
tract

popliteal fossa

tendons of
hamstrings

patella

patellar tendon

head of
fibula

peroneal
muscles

gastrocnemius

lateral
popliteal
nerve

extensors
of toes

subcutaneous
surface of tibia

soleus

achilles
tendon

medial
malleolus
(tibia)

lateral
malleolus
(fibula)

FIG. 2/13.—ANTERIOR AND POSTERIOR ASPECTS OF RIGHT LOWER LIMB

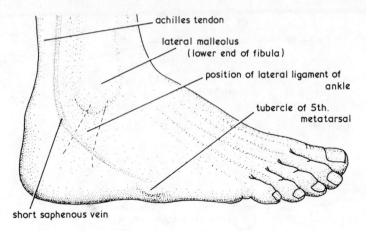

FIG. 2/14.—LATERAL ASPECT OF THE RIGHT FOOT

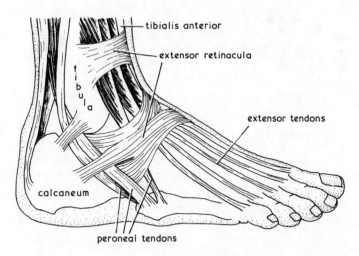

FIG. 2/15.—LATERAL ASPECT OF RIGHT FOOT, INDICATING THE POSITION OF THE TENDONS

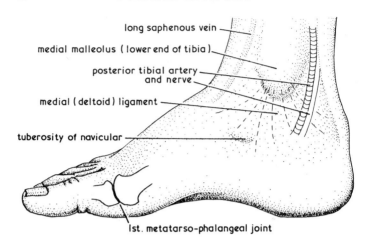

long saphenous vein

medial malleolus (lower end of tibia)

posterior tibial artery and nerve

medial (deltoid) ligament

tuberosity of navicular

1st. metatarso-phalangeal joint

FIG. 2/16.—THE MEDIAL ASPECT OF THE RIGHT FOOT

For Bones of Foot, *see* Fig. 6/8, page 108.

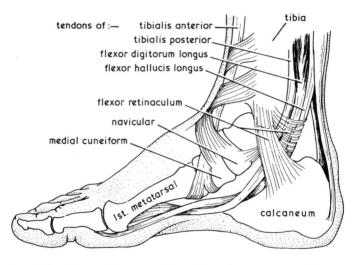

tendons of :—   tibialis anterior

tibialis posterior

flexor digitorum longus

flexor hallucis longus

tibia

flexor retinaculum

navicular

medial cuneiform

1st. metatarsal

calcaneum

FIG. 2/17.—MEDIAL ASPECT OF RIGHT FOOT INDICATING THE POSITION OF SOME OF THE TENDONS OF THE MUSCLES SUPPORTING THE ARCHES OF THE FOOT

Chapter 3

THE SKELETAL SYSTEM—BONES OF SKULL AND THORACIC CAGE

The Skeleton is the bony framework of the body *providing support and protection* for some of the soft organs, particularly in the skull and pelvis; *acting as levers* in movement and providing *surfaces for the attachment of the skeletal muscles.* In certain parts the framework is supplemented by cartilage.

The Axial Skeleton comprises the head and trunk, and includes the following bones:

Skull	Sternum and Ribs
Vertebral Column	Hyoid bone

The Appendicular Skeleton comprises the limbs, and limb girdles.

Upper limb
Lower limb

In addition, there are three small bones in each middle ear.

Classification of Bones. The bones of the skeleton are classified according to their shape and formation.

Long Bones. These are found principally in the limbs. Each long bone consists of a shaft and two extremities. Long bones act as levers in the body and make movement possible.

Short Bones. Good examples of these are seen in the bones of the carpus and tarsus. They are composed largely of cancellous bone tissue as they require to be light and strong. They have a thin covering of compact tissue. Short bones give strength in support, as in the strength shown in the wrist.

Flat Bones consist of two layers of dense bone tissue with an intervening layer of spongy bone. These are found where protection is needed, as in the bones of the skull, the innomin-

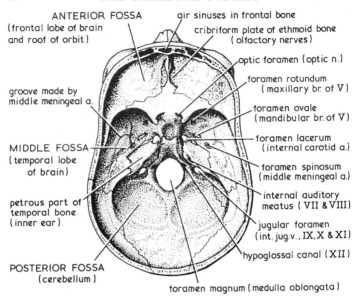

FIG. 3/1.—THE INTERIOR OF THE CAVITY OF THE SKULL, SHOWING THE ANTERIOR, MIDDLE AND POSTERIOR CRANIAL FOSSAE, THE FORAMINAE AND NERVES AND BLOOD VESSELS WHICH PASS THROUGH THEM

ate bones, ribs and scapulae. Flat bones also afford large surfaces for the attachment of muscles, e.g. the scapula.

Irregular Bones are those which cannot be included in either of the other three classes. Examples of irregular bones are the vertebrae and some of the bones of the face.

The Sesamoid Bones are another group. These are developed in the tendons of muscles, and are found in the vicinity of a joint. The patella is the largest example of this type.

THE SKULL

The skull is the bony framework of the head, arranged in two parts—the *cranium* (sometimes called the *calvaria*) consists of eight bones, and the *facial skeleton* of fourteen bones.

The cavity of the cranium presents an upper surface known

as the *vault of the skull*; this is smooth on its outer surface, and marked by ridges and depressions to accommodate the brain and its blood vessels on the inner surface. The lower

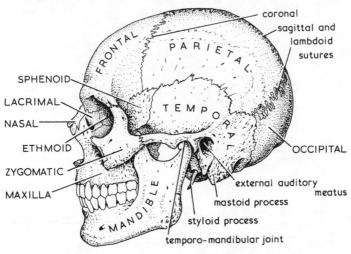

Fig. 3/2.—The Left Side of the Skull, Indicating the Position of Some of the Bones and Sutures and the Temporo-Mandibular Joint

surface of the cavity is known as the *base of the skull*. It is perforated by many holes for the passage of nerves and blood vessels.

Cranial Bones:

1 Occipital	2 Temporal
2 Parietal	1 Sphenoid
1 Frontal	1 Ethmoid

The Occipital Bone is at the back and lower part of the cranial cavity. It is pierced by the *foramen magnum*, through which the *medulla oblongata* passes to join the *spinal cord*. Each side of the foramen magnum are masses of bone which form the *condyles of the skull* and present articulating surfaces for the atlas (*see* Fig. 3/3).

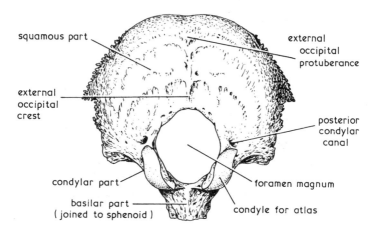

FIG. 3/3.—THE OCCIPITAL BONE VIEWED FROM BELOW

The two Parietal Bones together form the roof and sides of the skull. The outer surface is smooth, but the inner surface is marked by deep furrows which lodge the cranial arteries.

A very large furrow about the middle of the bone lodges the *middle meningeal* artery (*see* Fig. 3/1).

Rupture of this artery causes pressure on the soft brain tissue with subsequent damage, first on the same side and later on the opposite side as well. This results in alteration in the size of the pupils, a very important nursing observation during the care of patients with head injuries (*see* Clinical Note, page 73).

The Frontal Bone forms the forehead and the upper part of the orbital cavities. The *supraorbital margin* is marked by the supraorbital notch in its inner half; through this notch the supraorbital vessels and the supraorbital nerve pass. The inner surface of the frontal bone is marked by depressions, which are produced by the convolutions of the brain.

Two *Temporal Bones* form the lower part of the sides of the skull. Each bone consists of a number of parts.

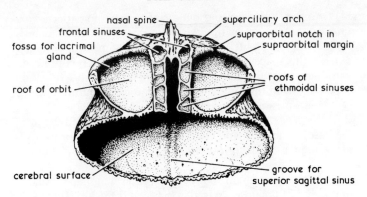

FIG. 3/4.—THE FRONTAL BONE SEEN FROM BELOW

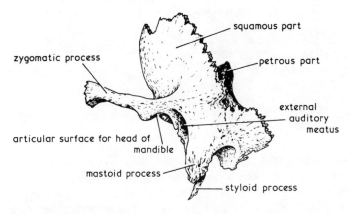

FIG. 3/5.—THE LEFT TEMPORAL BONE, SEEN FROM THE SIDE AND BELOW

The *squamous* part, *squama*, projects upwards and gives attachment to the temporalis muscle. From this the *zygomatic process* or *zygoma* projects forwards to join the zygomatic bone. Behind and below the root of this process lies the *external auditory meatus*.

The *mastoid portion* lies behind and it is continued downwards as the *mastoid process*; its outer surface gives attachment to the sternomastoid muscle. The mastoid process contains spaces known as the *mastoid air cells* and a particularly large space lying a little in front of these is named the *tympanic antrum*; this space is lined with epithelium which is continuous with that of the middle ear or tympanic cavity. Infection spreading from the middle ear may lead to suppuration in the tympanic antrum.

The *petrous* portion of the temporal bone is wedged in at the base of the skull and contains the hearing apparatus. (*See also* Chapter 26, section on Ear.)

The Ethmoid is a light spongy bone, cubical in shape, situated at the roof of the nose wedged in between the orbits. It consists

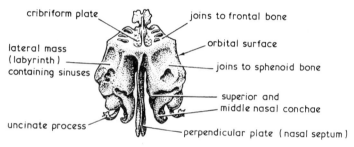

FIG. 3/6.—THE ETHMOID BONE SEEN FROM BEHIND

of two lateral masses or *labyrinths* composed of the ethmoidal cells or sinuses, which are closed except where they communicate with the nasal cavity. The ethmoid contains also a *perpendicular plate* and a *cribriform plate*. The perpendicular plate forms the upper part of the nasal septum. The cribriform plate fits into a notch on the frontal bone. Above this plate the olfactory bulbs lie, and through the perforations in the plate filaments of the olfactory nerves pass to the upper part of the nose (*see* Fig. 24/2, page 369).

The Sphenoid is similar in shape to a bat with wings outstretched, it consists of a body and two greater and two lesser wings. The body shows a depression named the *sella turcica* which lodges the *pituitary gland* or *hypophysis cerebri* (pituitary fossa) (*see* page 276).

It lies at the base of the skull and forms a large part of the middle cranial fossa (*see* Fig. 3/7).

The Sutures of the Cranium. The bones of the skull are united together by immovable joints called sutures, with the exception of one of the bones of the face, the mandible or lower jaw, which articulates with the temporal at the mandibular joint (*see* Fig. 3/2).

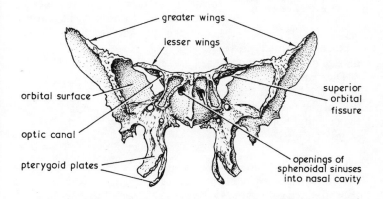

FIG. 3/7.—THE SPHENOID BONE SEEN FROM THE FRONT

The principal sutures are:

The *coronal* suture between the frontal bone and the two parietal bones.

The *sagittal* suture between the two parietal bones, running from before backwards along the top of the skull, and

The *lambdoid* suture between the occipital bone and the two parietal bones.

The Fontanelles. The bones of the skull of an infant at birth are not completely ossified. The spaces between the bones are filled in by membrane, and at the angles of the bones these membranes are called fontanelles. The largest of

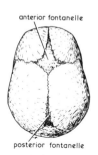

anterior fontanelle

posterior fontanelle

FIG. 3/8.—NEONATAL SKULL, VIEWED FROM ABOVE

these is situated at the junction of the frontal and the two parietal bones, where the coronal sagittal sutures meet. This is called the *anterior fontanelle*. It is diamond-shaped, measures about 4 cm ($1\frac{1}{2}$ inches) from back to front, and forms a soft spot on the head of an infant through which the brain can be felt pulsating. This fontanelle normally closes at the age of eighteen months.

The *posterior fontanelle* lies at the back, at the junction of the two parietal and the occipital bones. It closes soon after birth.

Air Sinuses of the Skull. Several cavities or chambers are contained in the bones of the skull. The *frontal, maxillary, ethmoid* (Fig. 3/9), and

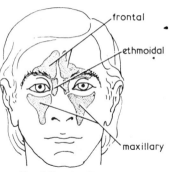

frontal

ethmoidal

maxillary

FIG. 3/9.—AIR SINUSES

sphenoid sinuses (Fig. 3/7) are the *paranasal* (accessory) sinuses which communicate with the nose. These air sinuses lighten the weight of the skull and give resonance to the voice.

The *frontal sinuses* lie in the frontal bone, one on each side at the root of the nose above the inner angle of the eye. The *maxillary sinuses*, sometimes known as the *antra of Highmore*, lie one on each side of the nose in the maxillary bones. (*See* Clinical Note, page 74.)

A number of small spaces known as the *mastoid cells* lie in the temporal bones; the *mastoid antrum* is the largest of these and lies in the mastoid process. It communicates with the tympanic cavity.

BONES OF THE FACE

There are 14 facial bones, all except the mandible being united by sutures and immovable.

Two *Nasal Bones* form the bridge of the nose.

Two *Palatine Bones* form the roof of the mouth and the floor of the nose.

Two *Lacrimal Bones* form the tear ducts and part of the orbit at the inner angle of the eye, through which the fluid from the eye is carried to the nasal cavity.

Two *Zygomatic Bones* form the cheek bones. Processes from these bones unite with the zygomatic processes of the temporal bones to form the *zygomatic arch.*

One *Vomer* forms the lower part of the bony partition in the nose. (The upper part of the nasal septum is formed by the perpendicular plate of the ethmoid.)

Two *Inferior Turbinate Bones* are the larger pair of three projections, *nasal conchae*, from the lateral wall of the maxilla.

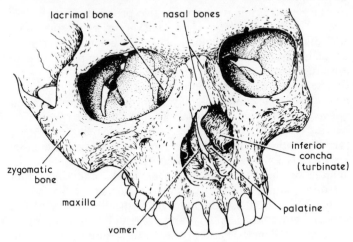

FIG. 3/10.—THE BONES OF THE FACE

Two *Maxillae* form the upper jaw and contain the upper teeth. The body of the maxilla contains a large cavity, the maxillary sinus, or antrum of Highmore, which communicates with the nasal cavity by two small openings. (*See* Fig. 3/9.)

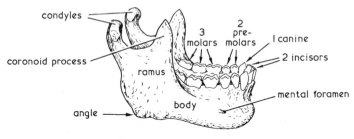

FIG. 3/11.—THE MANDIBLE

The *Mandible* forms the lower jaw. It is the only movable bone in the skull apart from the ossicles of the ear. It consists of a *body* which is the central curved horizontal part containing the lower teeth and forming the chin, and two upright portions called *rami*, one at each side, which join the *body* at the *angle of the jaw*.

The ramus terminates above in two processes, the *coronoid process* in front and the *condyle of the jaw*, or, as it is sometimes called, the *head* of the mandible which lies behind. This mandibular head or condyle articulates with the temporal bone to form the *temporomandibular joint*.

The *mandible* may be depressed and elevated as in opening and closing the mouth; it may be protruded, retracted and moved slightly from side to side as in mastication. (*See* p. 74.)

The Formation of the Nose. The bony framework of the nose, or the nasal fossae, is composed of two cavities about the middle of the face, separated from each other by a thin partition, which extends from the palate up to the frontal bone. These cavities communicate with the sinuses of the frontal, ethmoid, maxillary and sphenoid bones. Infection may spread from the nasal cavities to these sinuses.

For the general formation of the mouth, *see* page 213.

Clinical Notes

Head Injuries occur in from 50 to 80 per cent of all road and traffic accidents and no head injury can be lightly regarded. It is the commonest cause of all young male deaths. The *vault* or the base of the skull, or both, may be fractured.

First Aid is urgently needed. A state of *cerebral concussion* occurs almost immediately and may be so transient that it escapes notice; if seen the patient should be put almost prone (in coma position) with the head lowered, as the cough and the swallowing reflexes will be absent, so that any bleeding from the mouth or regurgitation of food or fluid from stomach or oesophagus runs out of the mouth and is not inhaled.

If not breathing the airway should be cleared and if necessary artificial respiration started until medical aid comes.

Observe bleeding. Scalp wounds bleed freely and can generally be controlled by a pad firmly bandaged (using a crepe bandage) over bleeding sites until help comes and the scalp can be sutured. Intracranial haemorrhage may be extradural and generally necessitates operation, or the bleeding may be subdural.

Levels of consciousness. A nurse should be familiar with these, noting the degree of responsiveness in order to be able to report indications of deterioration or improvement. She should also be familiar with the observation and care of an unconscious patient.

Discharge from ears and nostrils may be blood, but may also be cerebrospinal fluid escaping as the result of a fracture of the base of the skull.

The state of the pupils, if dilated or unequal, the condition of reflexes to light, the presence of squint, give valuable information to a surgeon. Necessary observations include routine estimation of blood pressure, temperature, pulse and respiration; the state of the skin as regards colour, warmth, sweating; restlessness, need for restraint of the patient, possibility of a distended bladder or need to empty the bowel, indications of pain or headache, noisiness of the patient, alteration and difficulty in speech, and difficulty in swallowing.

Damage to the contents of the skull. Any part of the brain may be injured. Reference to Fig. 22/8, page 337, indicates areas of the cerebrum which may be affected, giving rise to interference with movement (the motor cortex); with sensation (the sensory areas); impairment of the higher mental functions and emotions in the frontal lobe; vision in the occipital lobe; the speech mechanism, memory and hearing, the temporal lobe; injury to the internal capsule, resulting in disorders of movement and sensation. Any of the cranial nerves (*see* page 341) may be injured.

Increased intracranial pressure may be due to bleeding or to cerebral oedema. Either may give rise to:

(a) *cerebral compression* with loss of consciousness, a full-bounding pulse and hyperpyrexia, or

(b) *cerebral irritation* when the patient may be restless, disorientated and abusive.

Oedema may be treated by dehydrating drugs, such as urea, or mannitol hexanitrate.

Chest complications may occur, necessitating pharyngeal suction and in some cases tracheostomy.

Sequelae following injuries to the head are numerous and include:
Motor and sensory paralyses

Unsteadiness of gait
Traumatic epilepsy
Changes in personality

The *accessory air sinuses* (*see* Fig. 3/9, page 70) may be the site of infection, collectively termed *sinusitis*, spreading from the nose, with which they are intimately connected.

Infection of the frontal sinuses (frontal sinusitis) which are contained in the frontal bone, causes severe headache, a rise of temperature and malaise; these sinuses are closely related to the frontal lobe of the brain, giving rise occasionally to a frontal lobe abscess. A fracture of the base of the skull may involve these sinuses, resulting in leakage of cerebrospinal fluid should the dura mater be torn.

The maxillary sinus or *antrum of Highmore* occupies space in the maxilla; it may be infected from the nose or from infected teeth. There is throbbing pain over the cheek and similar constitutional symptoms as in *ethmoiditis*.

(For infection of the mastoid antrum, *mastoiditis*, *see* Clinical Note, Chapter 26, page 387).

The **mandible** may be fractured in an injury to the face, but more often it is dislocated forwards by a blow or even in yawning when the head of the mandible (*see* Figs. 3/2 and 3/11, pages 65, 72) slips forward.

To obtain a clear airway in a patient who is unconscious, all that may be required is to lift the jaw forwards by the fingers placed behind the angles of the jaw to prevent the tongue impinging on the posterior phalangeal wall and creating an obstruction. This can be practised on oneself and all nurses should be adept at this simple resuscitation measure, but great care must always be taken in dealing with a patient when the mandible is fractured.

THE THORACIC CAGE

The skeleton of the thorax is made up of bone and cartilage. The thorax is a cone-shaped cavity, broader below than above, and longer behind than in front.

It is formed by the *twelve thoracic vertebrae at the back*, the *sternum in front*, and the *twelve pairs of ribs at the sides*, which encircle the trunk from the vertebral column behind to the sternum in front. (To avoid repetition *see* Thoracic cavity, page 256 and fig. of skeleton at the end of the book.)

The Sternum. The sternum or breast-bone is a flat bone divided into three parts.

The Manubrium Sterni is a triangular-shaped piece of bone placed above the body of the sternum. It articulates with the clavicles at its upper and outer aspect on each side by means of the *clavicular notches*. These articulations are separated by the *suprasternal* or *jugular notch*.

The first pair of ribs articulates with the sides of the manubrium, and the second pair at the junction of the manubrium and the body of the sternum. The joint between the manubrium sterni and the *gladiolus* or body of the sternum is a symphysis; a pad of cartilage separates the joint surfaces. This junction is called the *angle of Ludwig* or *angle of Louis*; it corresponds in position to the level of the second rib.

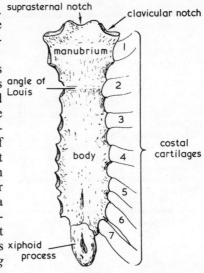

FIG. 3/12.—THE ANTERIOR ASPECT OF THE STERNUM

The Body of the Sternum is long and narrow and notched on each side for the attachment of the third, fourth, fifth, sixth, and seventh costal cartilages.

The Ensiform Process or *Xiphoid Bone* is the lowest part of the sternum. It is cartilaginous in youth but ossifies in older subjects. The diaphragm, the linea alba, and the rectus abdominis are attached to the *Xiphoid Bone*.

The Ribs. There are twelve pairs of ribs. They are attached behind to the thoracic vertebrae, articulating with them by means of facets on the sides of the bodies of the vertebrae, and on the transverse processes, which correspond with similar facets on the head of each rib.

The upper seven pairs of ribs are attached to the sternum anteriorly by means of their *costal cartilages*. These are the *true ribs*. *The first rib is the shortest,* and the subclavian vein, the subclavian artery and the lowest trunk of the brachial plexus pass over this rib (*see* page 144). In the presence of a

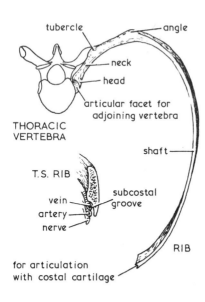

THORACIC
VERTEBRA

tubercle

angle

neck

head

articular facet for
adjoining vertebra

shaft

T.S. RIB

subcostal
groove

vein

artery

nerve

RIB

for articulation
with costal cartilage

FIG. 3/13.—A TYPICAL RIB AND VERTEBRA
SEEN FROM ABOVE

cervical rib, which is an abnormality, these vessels and nerves may be subject to pressure, resulting in effects such as interference with the blood supply to the hand and a pins and needles feeling in the fingers.

Of the lower five pairs of ribs, the eighth, ninth and tenth are attached indirectly to the sternum by means of the attachment of their costal cartilages to the cartilage of the rib above. *The last two pairs of ribs* are unattached in front and are called *floating ribs* (*see* fig. of skeleton at end of book).

A rib is described as a long bone. It has two extremities, anterior and posterior, and a shaft. The vertebral or posterior extremity of the rib presents a head, a neck, and a tubercle. The anterior or sternal end has a depression for the attachment of the costal cartilage. The shaft is thin and flat, it has an inner and an outer surface and an upper and lower border; the internal surface is smooth and is marked by a groove, the *subcostal groove* in which lie the intercostal vessels and nerve. Thus a needle passed into the chest, *above a rib,* will avoid injury to these structures.

The ribs slope downwards from back to front. The posterior extremity of the rib is the more fixed point and the anterior end the more movable. By reason of the elasticity of the costal cartilages the movements of the ribs in respiration are very free.

The Costal Cartilages are bars of hyaline cartilage which connect the ribs with the sternum, and by means of their elasticity allow of considerable movement. The cartilages attached to the two last ribs are pointed.

The Intercostal Spaces vary in size between each two ribs, and are filled in by the *intercostal muscles* which lie between them, thus closing the spaces and helping to form the thoracic cavity (*see* page 256).

The principal groups are the *external intercostal muscles* arising from the lower border of the rib above and passing to the upper border of the rib below, the fibres running obliquely downwards and forwards. The *internal intercostal muscles* occupy the same position but their fibres run obliquely downwards and backwards. (*See* Clinical Note, below, on stress fracture of the ribs.)

Clinical Notes

The elastic costal cartilages (above) serve to protect the sternum from injury, although a fracture of the dorsal spine may result in a sternal fracture by indirect violence.

Sternal puncture is made by a wide bore needle passed into the vascular cancellous layer of bone to obtain, by aspiration, a specimen of bone marrow, in order to examine the condition of the red blood cells which are developing in the marrow.

The sternum is divided to gain access to the mediastinum and also for operation on the heart.

Fracture of the Ribs. In a child, as the ribs are elastic, fracture is rare; in adults fractures do occur, generally as the result of indirect violence, when the break occurs at the angle of the rib. As a rule, breathing exercises are taught, the patient is encouraged to move about and the ribs heal. Discomfort is experienced for about a week.

When the chest is severely crushed, the ribs may be driven in and cause injury to an organ lying immediately beneath, either in the thorax or in the abdominal cavity. In this case a firm application of strapping is applied to limit movement of the fragments.

Stress fracture of the ribs may occur in debilitated persons due to over-action of the intercostal muscles. It may occur in coughing.

A note on cervical rib is given on page 76.

In *external cardiac massage,* unless expertly given, costal cartilages may be dislocated and displaced and ribs broken.

Chapter 4

THE VERTEBRAL COLUMN AND PELVIC GIRDLE

The vertebral or spinal column is a flexible structure formed by a number of bones called vertebrae. Between each two bones the column is supplemented by pads of fibro-cartilage. The adult column measures 60–70 cm (24–28 inches) in length. There are 33 vertebral bones, 24 of these are separate bones and the remaining 9 vertebrae are fused to form 2 bones.

The vertebrae are grouped and named according to the region they occupy.

Seven *cervical vertebrae* form the neck or cervical region.

Twelve *thoracic vertebrae* form the back of the thorax or chest.

Five *lumbar vertebrae* form the lumbar region or loins.

Five *sacral vertebrae* form the sacrum.

Four *coccygeal vertebrae* form the coccyx or tail.

The vertebrae in the three upper regions remain separate or distinct throughout life, and are called *movable vertebrae*. Those in the two lower regions, the sacrum and coccyx, are united in the adult to form two bones, the *fixed vertebrae*.

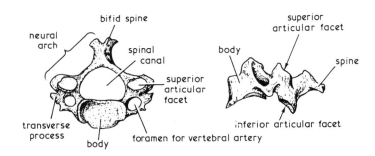

Fig. 4/1.—A Typical Cervical Vertebra, from Above and the Side

With the exception of the first two cervical vertebrae, all the movable vertebrae have similar characteristics; a typical vertebra consists of two parts, an anterior part called the *body*, and a posterior part called the *neural arch* enclosing the neural canal through which the spinal cord passes (*see* Figs. 4/1 to 4/4).

The Cervical Vertebrae are the smallest of the bones, and except the *first* and *second*, which are peculiar in shape (*see* Fig. 4/2), the cervical vertebrae possess the following characters in common. The bodies are small and oblong in shape, broader from side to side than from before backwards. The neural arch is large. The spinous processes are divided or bifid terminally. The transverse processes are perforated by foramina for the passage of the vertebral arteries.

The *seventh cervical vertebra* is the first vertebra with an undivided spinous process. This process has a tubercle at its tip. It forms a distinct projection in the neck and can be seen at the lower part of the back of the neck. Because of this characteristic the bone is called the *vertebra prominens*.

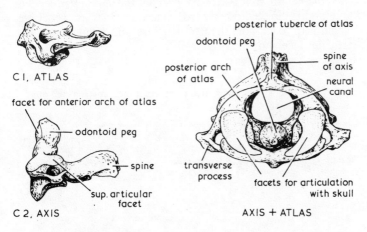

FIG. 4/2.—THE AXIS PRESENTS THE ODONTOID PEG AROUND WHICH THE ATLAS MOVES IN THE ROTATION OF THE HEAD

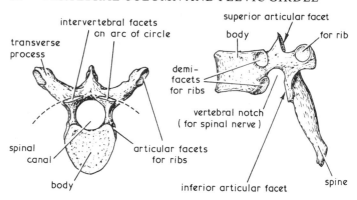

FIG. 4/3.—A TYPICAL THORACIC VERTEBRA, FROM ABOVE AND THE SIDE

A rib articulates with the transverse process and superior demifacet and the inferior demifacet of the vertebra above.

The Thoracic Vertebrae are larger than the cervical and they increase in size as they extend downwards. A typical thoracic vertebra presents the following characteristics. The body is heart-shaped, with facets on each side for attachment of the ribs, the neural arch is relatively small, the spinous process is long and is directed downwards, and the transverse processes which help to support the ribs are thick and strong and carry articular facets for the ribs (*see* Fig. 4/3).

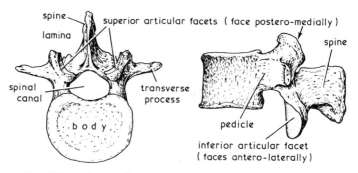

FIG. 4/4.—A TYPICAL LUMBAR VERTEBRA, FROM ABOVE AND THE SIDE

The Lumbar Vertebrae are the largest. The body is very large compared with the bodies of the other vertebrae and is kidney-shaped. The spinous process is broad and hatchet-shaped. The transverse processes are long and slender. The fifth lumbar vertebra articulates with the sacrum at the lumbo-sacral joint.

The Sacrum is a triangular bone situated at the lower part of the vertebral column, wedged in between the two innominate bones, and forming the back of the pelvic cavity. The *base of the sacrum* lies above and articulates with the fifth lumbar vertebra, forming a typical intervertebral joint. The anterior edge at the base of the sacrum forms the *sacral promontory*. The *sacral canal* lies below the spinal canal and is a continuation of it. The walls of the sacral canal are perforated for the passage of the sacral nerves. Rudimentary spinous processes can be seen on the posterior aspect of the sacrum. The *anterior surface of the sacrum* is concave and shows four

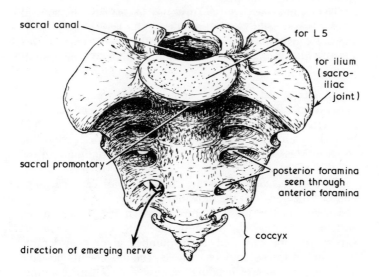

sacral canal

for L 5

for ilium (sacro-iliac joint)

sacral promontory

posterior foramina seen through anterior foramina

coccyx

direction of emerging nerve

FIG. 4/5.—THE ANTERIOR SURFACE OF THE SACRUM AND COCCYX

transverse ridges, which mark the points of union of the five sacral vertebrae. At the extremities of these ridges, on each side, are apertures for the passage of nerves. These are called the *sacral foramina*. The *apex of the sacrum* articulates with the coccyx. At the sides the sacrum articulates with the innominate bones, forming the right and left sacro-iliac joints.

The Coccyx is composed of four or five rudimentary vertebrae, fused to form one bone. It articulates above with the sacrum.

The Curves of the Vertebral Column. Looked at from the side, the vertebral column presents four antero-posterior curves; the *cervical curve* in the neck which is convex forwards, the *thoracic curve,* convex backwards, the *lumbar curve,* convex forwards, and the *pelvic curve,* convex backwards.

The two posteriorly convex curves, thoracic and pelvic, are called *primary* as they persist from the total backward convexity of the spine which is C-shaped *in utero* with the head bent downwards on to the chest and the pelvic girdle tilted upwards towards the body.

The two anteriorly convex curves are *secondary*—the cervical curve is developed when an infant raises his head to look about and investigate his surroundings, and the lumbar curve forms when he crawls and learns to stand and walk and keep himself erect (*see* Fig. 4/6).

The Joints of the Vertebral Arches. These are cartilaginous joints formed by pads of fibro-cartilage placed between each two vertebrae, strengthened by ligaments running in front and behind the vertebral bodies throughout the entire length of the column. Masses of muscle on each side materially aid in the stability of the spine.

The Intervertebral Discs are thick pads of fibro-cartilage between the bodies of the movable vertebrae.

Movement. The joints formed between the discs and the vertebrae are only slightly movable joints of the *symphysis* variety, *see* p. 113, but their number gives considerable flexibility to the column as a whole. The movements possible are—*flexion,* forward bending, *extension,* backward bending, *lateral bending* to each side and *rotation* to right and left.

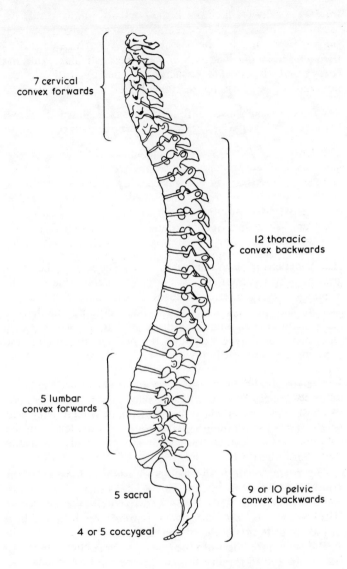

7 cervical
convex forwards

12 thoracic
convex backwards

5 lumbar
convex forwards

5 sacral

4 or 5 coccygeal

9 or 10 pelvic
convex backwards

FIG. 4/6.—THE CURVES OF THE VERTEBRAL COLUMN

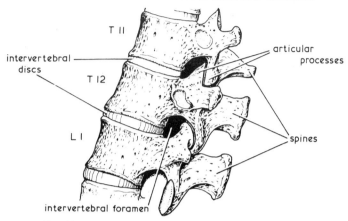

FIG. 4/7.—THE LATERAL ASPECT OF INTERVERTEBRAL JOINTS

The Functions of the Vertebral Column. The vertebral column acts as a firm support to the body, yet at the same time, by means of its intervertebral discs of cartilage which act as buffers, and its curves which give flexibility and enable it to bend without breaking, these discs serve to absorb shock set up when moving the weight of the body as in running and jumping, and thus the brain and spinal cord are protected from shocks and jars (*see* Figs. 4/6 and 4/7).

The vertebral column also supports the weight of the body, affords surfaces for the attachment of muscles, and forms a strong posterior boundary for the cavities of the trunk and gives attachment to the ribs.

THE PELVIC GIRDLE OR BONY PELVIS

The pelvic girdle is the means of connexion between the trunk and lower extremities. This girdle is formed by part of the axial skeleton—the *sacrum* and *coccyx* being wedged in between the two *innominate bones*. It articulates with its fellow of the opposite side at the symphysis pubis.

The pelvis is divided into the *true pelvis* or pelvic basin which lies below the brim, and the *false pelvis* formed by the iliac bones extending above the brim. The *inlet* of the true pelvis is

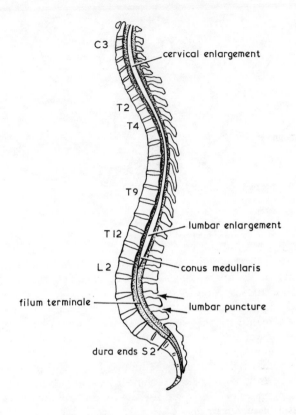

FIG. 4/8.—SPINAL CORD IN RELATION TO THE VERTEBRAL COLUMN

The *Cervical enlargement* of the spinal cord extends from the third cervical to the second thoracic vertebra. The *Lumbar enlargement* begins at about the level of the ninth thoracic, and below the level of the twelfth thoracic vertebra it tapers to form the *conus medullaris* and ends at the lower border of the first lumbar vertebra or about the upper border of the second. When lumbar puncture is performed a needle is passed into the sub-arachnoid space through the interval between the third and fourth or the fourth and fifth lumbar vertebrae, thus avoiding possible injury to the spinal cord.

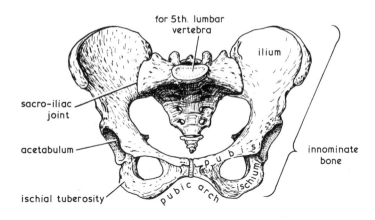

FIG. 4/9.—THE FEMALE PELVIS

The female pelvis, adapted for childbirth, is wide and shallow, the inlet is large and round, the pubic arch is wide, the ischial tuberosities are farther apart than in the male and the coccyx is movable.

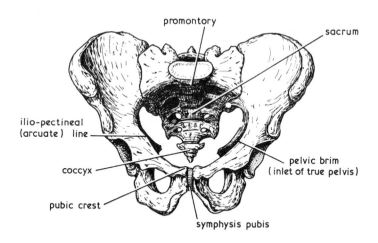

FIG. 4/10.—THE MALE PELVIC GIRDLE

The male pelvis is longer and narrower than the female. The bones are stronger, the muscular attachments are well marked, the inlet is smaller and it is heart-shaped.

the brim, formed by the promontory of the sacrum, the ilio-pectineal line (on each side), and the crest of the pubic bones. The *outlet* is bounded by the coccyx and the ischial tuberosities.

The Joints of the Pelvis. The *sacro-iliac joint* is an articulation between the articular surfaces of the ilium (which are called *auricular* because of their similarity in shape to the auricle of the ear), and the sides of the sacrum. Only slight movement is possible at this joint as very strong ligaments unite the articulating surfaces, limiting movement in all directions.

The *symphysis pubis* is a cartilaginous joint between the pubic bones, which are separated by a pad of cartilage.

Clinical Notes

The curves of the vertebral column. The skeleton at the end of the book (side view) shows the bony framework in good postural standing position and the vertebral column (Fig. 4/6), the antero-posterior curves well balanced. An exaggeration of the backward thoracic convexity results in round back or *kyphosis*. This round back is a cause of poor thoracic expansion, often associated with disease of the chest, such as bronchitis. The head is poking forward and the chest flat. In exaggeration of the forward lumbar convexity, hollow back or *lordosis*, the pelvis is tilted forwards, the abdominal muscles relaxed, and strain is thrown on the ligaments at the front of the hip joint (*see* page 122). In both kyphosis and lordosis flat foot may result (*see* page 111).

The intervertebral discs shown in Fig. 4/7 may be affected by accident or age. Each fibro-cartilage or disc has a jelly-like centre or nucleus contained in a fibrous capsule. Prolapse of this nucleus through the capsule—*prolapsed intervertebral disc*—may cause pressure on adjacent nerve roots leading to pain and sometimes loss of power in the area of distribution of the nerve or nerves affected. A prolapsed lumbar disc is a common cause of sciatica. Disc shrinkage with degenerative changes occur in ageing.

The spine may be fractured either by direct violence as in a severe crushing accident or indirectly as when a weight falls on head and shoulders and the spine, unable to adapt to the weight, snaps. The commonest accident results in a fracture dislocation and in this the spinal cord may be badly injured between the displaced vertebrae.

The *symptoms* are those described in transection of the spinal cord (*see* page 357).

The pelvis also may be fractured and, if in two places, over-riding of the fragments may result in injury to some of the pelvic organs. It may be *contracted* in the case of a small woman, thus narrowing the inlet; in the *rachitic flat pelvis,* rarely seen in this country, the inlet is considerably decreased in diameter, making the mechanism of childbirth difficult or impossible.

Chapter 5

THE SKELETON OF THE UPPER LIMB

The skeleton of the upper limb is attached to the skeleton of the trunk by means of the *shoulder girdle,* which consists of the *clavicle* and *scapula.*

Below this the following bones form the skeleton of arm, forearm, and hand, making altogether 30 bones:

Humerus	5 Metacarpals	8 Carpal bones
Ulna and Radius	14 Phalanges	

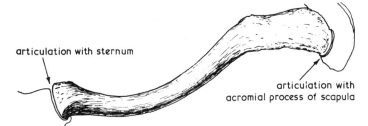

articulation with sternum

articulation with acromial process of scapula

FIG. 5/1.—THE UPPER SURFACE OF THE LEFT CLAVICLE

The Clavicle or collar bone is a long curved bone forming the anterior part of the shoulder girdle. It presents for examination a shaft and two extremities. The medial extremity is called the *sternal extremity* and articulates with the sternum. The lateral, the *acromial extremity*, articulates with the acromion process of the scapula.

Functions. The clavicle provides attachment for some of the muscles of the neck and shoulder and thus acts as a prop to the arm. (*See* Clinical Note on fracture of the clavicle, p. 97.)

SCAPULA

The scapula forms the posterior part of the shoulder girdle and lies at the back of the thorax superficially to the ribs. It is a triangular flat bone presenting for examination two surfaces, three angles, and three borders (upper, lateral and medial).

The Surfaces of the Scapula. The anterior or costal surface is called the *subscapular fossa*, and lies nearest the ribs. The posterior or dorsal surface is divided by a prominent ridge of bone, called the *spine of the scapula*, which passes across it to end in the *acromion process*, which overhangs the shoulder joint (*see* Fig. 5/3).

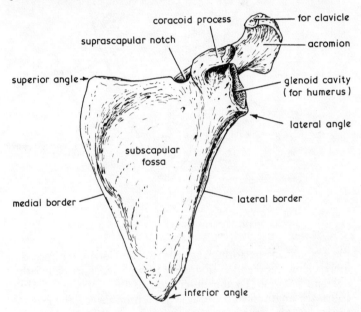

FIG. 5/2.—THE COSTAL ASPECT OF THE LEFT SCAPULA

HUMERUS

The humerus is a long bone, the longest bone of the upper limb. It presents a shaft and two extremities.

The Upper Extremity of the Humerus consists of one-third of a sphere—the *head*, which articulates with the glenoid cavity of

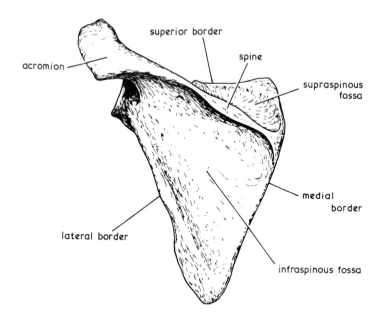

Fig. 5/3.—The Posterior Aspect of the Left Scapula, showing the Position of the Spine of the Scapula and the Borders

the scapula in the formation of the shoulder joint. Immediately below the head is a slightly constricted part called the *anatomical neck*. To the outer side of the upper extremity, below the anatomical neck, is a rough prominence, the *greater tuberosity*, and at the front is a smaller prominence, the *lesser tuberosity*. Between these tuberosities is a groove, the *bicipital groove* or *intertubercular sulcus*, in which the tendon of the biceps muscle lies. The bone becomes narrower below the tuberosities, and at this point it is called the *surgical neck*, because of the liability of fracture at that part (*see* Fig. 5/4).

The Shaft is rounded in its upper part, but becomes flattened from side to side as it approaches the lower extremity. A rough

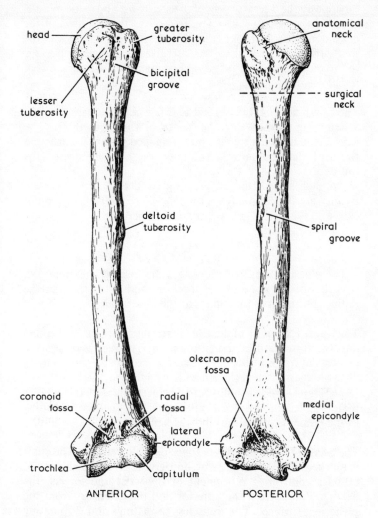

FIG. 5/4.—THE LEFT HUMERUS, SHOWING THE SALIENT POINTS MENTIONED IN THE TEXT

tubercle on the lateral aspect of the shaft, just above the middle, is called the *deltoid tuberosity*. It receives the insertion of the deltoid muscle. A groove runs obliquely across the back of the shaft, from the medial to the lateral aspect. It gives passage to the radial or musculo-spiral nerve and is called the *spiral* or *radial groove* (*see* Fig. 5/4).

The Lower Extremity is broad and flat. At its lowest part the articulating surfaces for the bones of the forearm lie. The *trochlea* on the inner side is a pulley-shaped surface for articulation with the ulna, and the *capitulum* on the outer side for the radius.

On each side of the articulating surfaces of the lower extremity are two *epicondyles*, the *lateral epicondyle* to the outer side, and the *medial epicondyle* to the inner side.

ULNA

The ulna is a long bone having a shaft and two extremities. It is the medial bone of the forearm, and is longer than the radius. The head of the ulna is at the lower end.

The Upper Extremity of the Ulna is strong and thick, and enters into the formation of the elbow joint. The *olecranon process* projects upwards at the back, and fits into the olecranon fossa of the humerus, when the elbow is straight.

The coronoid process of the ulna projects in front. It is smaller than the olecranon process, and it fits into the coronoid fossa of the humerus, when the elbow is bent.

The Shaft of the Ulna tapers towards its lower end. The shaft is marked off into surfaces by borders. It gives attachment to muscles controlling movement of the wrist and fingers, the flexors coming from the anterior and the extensors from the posterior surface. The muscles pronating and supinating the forearm are also attached to the shaft.

The Lower Extremity is small. Two eminences arise from it. A small rounded eminence, the *head of the ulna*, articulates

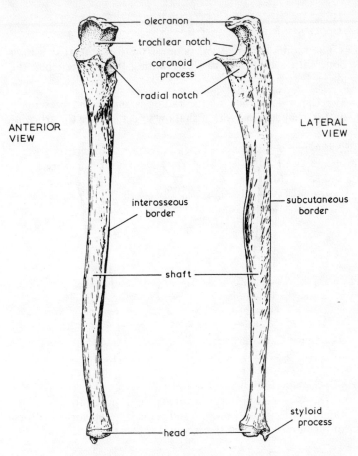

olecranon

trochlear notch

coronoid process

radial notch

ANTERIOR VIEW

LATERAL VIEW

interosseous border

subcutaneous border

shaft

styloid process

head

FIG. 5/5.—THE LEFT ULNA SHOWING ANTERIOR AND LATERAL ASPECTS, WITH SALIENT FEATURES

with the medial side of the lower extremity of the radius in the formation of the inferior radio-ulnar joint. A pointed process, the *styloid process*, projects downwards from the back of the lower extremity.

RADIUS

The radius is the lateral bone of the forearm. It is a long bone with a shaft and two extremities. It is shorter than the ulna.

The Upper Extremity of the radius is small, and presents a button-shaped head with a shallow upper surface for articula-

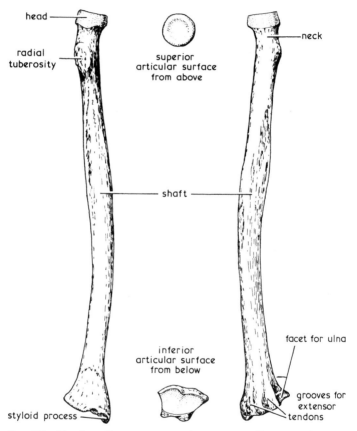

Fig. 5/6.—THE LEFT RADIUS SHOWING ANTERIOR AND POSTERIOR ASPECTS, AND SHOWING SALIENT POINTS

tion with the *capitulum* of the humerus; the sides of the head articulate with the *radial notch* of the ulna. Below the head lies the neck, and below and to the medial side of the neck lies the *radial tuberosity*, to which the tendon of insertion of the biceps muscle is attached.

The Shaft is narrower and more rounded above than below, widening as it nears the lower end. The shaft is curved outwards, and it is marked off into surfaces which, as in the ulna, give attachment anteriorly to the deep flexors and pronators, and posteriorly to the deep extensors and supinators of the forearm and hand. The interosseous ligament passes from radius to ulna and separates the muscles on the back from those on the front of the forearm.

The Lower Extremity is rather square in shape, and enters into the formation of two joints. The inferior articulating surfaces of the lower extremity of the radius articulate with the scaphoid and semilunar (lunate) bones in the formation of the wrist joint. An articulating surface on the medial aspect of the lower extremity articulates with the head of the ulna in the formation of the *inferior radio-ulnar joint*. The lateral aspect of the lower extremity is prolonged downwards into the *styloid process* of the radius.

BONES OF WRIST AND HAND

The bones of the hand are arranged in groups. The *carpus*, or the bones which enter into the formation of the wrist, are short bones. The *metacarpals* form the skeleton of the palm of the hand and are long bones. The *phalanges*, or bones of the fingers, are long bones.

The Carpus is composed of eight bones arranged in two rows, four bones in each row. The upper row is arranged from without inwards, navicular (scaphoid), lunate (semi-lunar), triquetral and pisiform.

The lower row, trapezium, trapezoid, capitate and hamate.

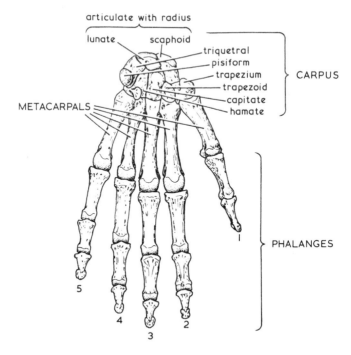

FIG. 5/7.—THE ANTERIOR ASPECT OF THE BONES OF THE LEFT WRIST AND HAND, GIVING THE NAMES AND RELATIVE POSITIONS OF THE INDIVIDUAL BONES

The *navicular* (*scaphoid*) is a boat-shaped bone; the *lunate* (*semi-lunar*) is crescentic like a halfmoon—these two bones articulate above with the lower extremity of the radius in the formation of the wrist joint, and below they articulate with some of the carpal bones of the second row.

The Metacarpus. There are five metacarpal bones. Each bone has a shaft and two extremities. The extremity articulating with the carpal bones is called the *carpal extremity*, and the joint so formed is the *carpo-metacarpal joint*. The *distal extremity* articulates with the phalanges and is called the head. The shafts of these bones are prismoidal, and have their broadest surface directed posteriorly (towards the back of the hand). The interosseous muscles are attached to the sides of the shafts.

The Phalanges are also long bones, having a shaft and two extremities. The shaft tapers towards the distal end. There are fourteen phalanges, three in each finger and two in the thumb.

Clinical Notes

The clavicle is the most commonly fractured bone in the body. It may be broken by direct or indirect violence, such as falling on the hand or the shoulder. The bone is usually fractured in the middle or medial third. The deformity is characteristic. The patient presents supporting the sagging limb with his opposite hand and forearm.

Fractures of the humerus are also common. The shaft is broken below the deltoid insertion, when the radial nerve may be injured. Fractures of the 'surgical neck', just below the head, are usually impacted, and the axillary (circumflex) nerve may be involved. In fracture of the lower end, particularly when the internal epicondyle is involved, the ulnar nerve may be injured. *Supracondylar fractures* of the humerus are common in children.

Either bone of the *forearm* may be broken. *Colles's fracture* is a transverse break of the lower end of the radius, about an inch above the wrist, common in elderly people by falling on to the outstretched hand; the ligaments are stretched and torn and the styloid process of the ulna may be fractured.
Displacement of the lower fragment of the radius upwards gives the unsightly 'dinner fork' deformity, which makes reduction and resetting essential before the bone can heal in good alignment.

Any of the *carpal bones*, most often the navicular, may be fractured. Carpal bones may be dislocated by falling heavily on the hand. Fractures of the metacarpals and phalanges are usually the result of direct violence.

The '*carpal tunnel syndrome*'. Along with the flexor tendons to the hand, the median nerve passes beneath the flexor retinaculum (*see* Fig. 8/11, page 138). Any condition (or there may not be any apparent cause) reducing the size of this 'tunnel' may give rise to pressure on the median nerve resulting in numbness, tingling and weakness of the muscles supplied by it.

Chapter 6

THE SKELETON OF THE LOWER LIMB

The bones of the lower extremity are connected with the trunk by means of the pelvic girdle, which is described on page 84.

The lower extremity consists of thirty-one bones:

1 Innominate bone	1 Patella
1 Femur	7 Tarsal bones
1 Tibia	5 Metatarsal bones
1 Fibula	14 Phalanges

THE INNOMINATE BONE

The innominate bone or *os innominatum* helps to form the pelvic girdle. Situated one on each side, uniting in front at the symphysis pubis, the two bones form a considerable part of the bony pelvis.

The innominate bone is an irregular flat bone formed by the union of three bones at the acetabulum, which is a cup-shaped cavity on the external surface of the bone which receives the head of the femur in the formation of the hip joint.

The uppermost of the three bones which unite here is the *ilium*, the front one the *pubis*, and the most posterior the *ischium*.

The Ilium presents two surfaces, a crest, and an articulating surface for the sacrum.

The *crest of the ilium* is curved and surmounts the bone. It gives attachment to many muscles, including the abdominal muscles and latissimus dorsi. It terminates in front at a point called the *anterior superior iliac spine*, to which the *inguinal ligament* (Poupart's) (*see* page 135) is attached, and posteriorly in the *posterior superior iliac spine*. Below these spines are two other prominences, the anterior and posterior inferior spines. The surface between the two posterior spines forms the articulating surface for the sacrum. Below this articulation lies the *greater sciatic notch*, through which the sciatic nerve passes from the pelvis to the thigh.

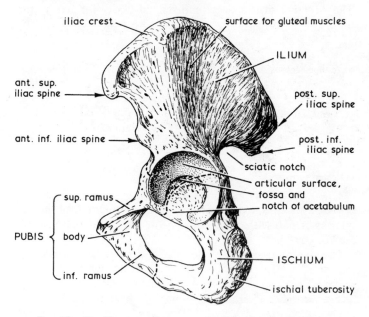

FIG. 6/1.—THE EXTERNAL SURFACE OF THE LEFT INNOMINATE BONE
Dotted lines indicate limits of component parts.

The Pubis consists of a *body* and two *rami*. The body is square in shape and is surmounted by the *crest of the pubis*. The pubic bones unite in front at the *symphysis pubis* (*see* Figs. 4/9 and 4/10).

The Ischium is the thickest and strongest portion of the bone. The *tuberosity of the ischium* lies at its lowest point, and on this the trunk rests when sitting. A pointed eminence, the *spine of the ischium*, arises from the back of the bone and marks the lowest part of the sciatic notch (*see* Fig. 6/2).

The Obturator Foramen is a large oval foramen lying below the acetabulum bounded by the pubis and ischium. It is filled in with membrane, and at its upper part the obturator vessels and nerves pass from the pelvis into the thigh.

The Acetabulum is a deep, cup-shaped cavity formed by the union of the three bones; the *pubis* forms the front part, the

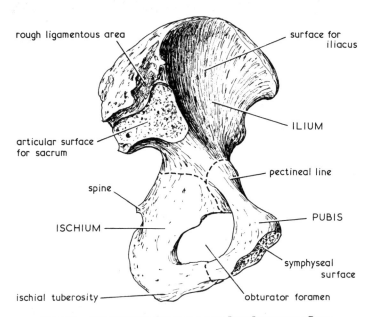

rough ligamentous area

surface for iliacus

ILIUM

articular surface for sacrum

pectineal line

spine

PUBIS

ISCHIUM

symphyseal surface

ischial tuberosity

obturator foramen

FIG. 6/2.—THE INTERNAL SURFACE OF THE LEFT INNOMINATE BONE

ilium the upper part, and the *ischium* the back part. The acetabulum articulates with the femur in the formation of the hip joint. The articulating surface is shaped like a horse-shoe interrupted at its lowest point by a notch, the *acetabular notch*, permitting the passage of vessels into the joint. A roughened non-articular surface at the bottom—*the acetabular fossa*—is filled in with a pad of fat; its lower margins give attachment to the *ligamentum teres* of the hip joint.

FEMUR

The femur is the longest bone in the body. It articulates with the acetabulum in the formation of the hip joint, and from here the bone inclines medially to the knee, where it articulates with the tibia. It is a long bone with a shaft and two extremities.

The Upper Extremity presents a head which forms two-thirds of a sphere; at the summit of this is an ovoid depression, a roughened pit, for the attachment of the ligamentum teres. Below the head is the neck, which is long and flattened. Where the neck joins the shaft the *greater trochanter* lies to the outer side, and the *lesser trochanter* to the back and inner side.

At the base of the neck of the bone, two lines unite the greater and lesser trochanters, the *intertrochanteric line* in front (*see* Fig. 6/3), and the *intertrochanteric crest* at the back (*see* Fig. 6/4). The latter is marked by a tubercle of bone, the *quadrate tubercle* half way along its length.

The Shaft of the Femur is cylindrical, smooth, and rounded in front, and at the sides. It curves forwards and has a very well marked ridge behind, called the *linea aspera*, to which a number of muscles are attached, amongst them the adductors of the thigh.

The Lower Extremity is wide and presents two condyles, an intercondylar notch, a popliteal surface, and a patellar surface. The condyles are very prominent; the medial one is lower than the lateral one. They both enter into the formation of the knee joint.

The *intercondylar notch* separates the condyles behind. The surfaces of this notch give attachment to the cruciate ligaments of the knee joint. The condyles are separated in front by the *patellar surface* which extends over the anterior aspect of both condyles; on this surface the patella rests. The tibial surface of the femoral condyles lies below and rests on the upper articulating surface of the condyles of the tibia. This surface is divided into two areas by the deep intercondylar notch. The *popliteal surface* of the bone lies above the condyles at the back.

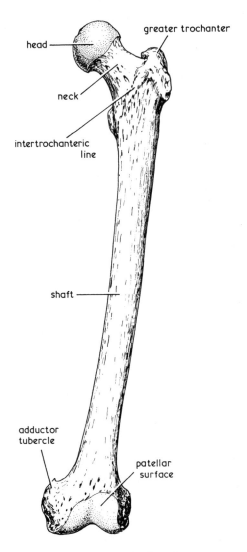

FIG. 6/3.—THE ANTERIOR ASPECT OF THE LEFT FEMUR, SHOWING THE SALIENT
FEATURES MENTIONED IN THE TEXT

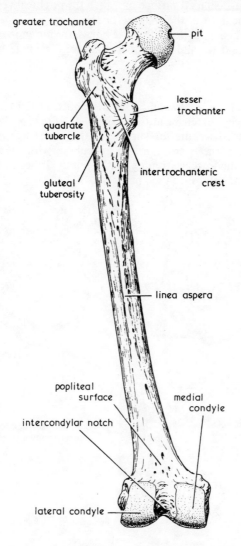

FIG. 6/4.—THE POSTERIOR ASPECT OF THE LEFT FEMUR, SHOWING THE SALIENT FEATURES MENTIONED IN THE TEXT

It is a lozenge-shaped surface on which the popliteal vessels lie. It forms the floor of the *popliteal* space.

The femur articulates with three bones, the innominate bone, the tibia, and the patella, but it does *not* articulate with the fibula.

PATELLA

The patella is a *sesamoid bone* developed in the tendon of the quadriceps extensor muscle. The *apex of the patella* points downwards. The anterior surface of the bone is rough. The posterior surface is smooth and articulates with the patellar surface of the lower extremity of the femur. It lies in front of but does *not* enter into the knee joint.

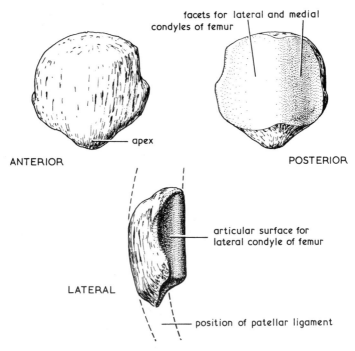

Fig. 6/5.—The Anterior, Posterior and Lateral Aspects of the Left Patella

TIBIA

The tibia or shin bone forms the main skeleton of the leg and lies medial to the fibula; it is a long bone with a shaft and two extremities.

The Upper Extremity presents medial and lateral condyles.
The *condyles* form the upper and most expanded portion of the bone. The superior surface of these presents the two plateau-like articulating surfaces for the femur in the formation of the knee joint. These surfaces are smooth, and on their flat surfaces lie the *semilunar cartilages*, which deepen the articulating surfaces for the reception of the femoral condyles. (*See* Fig. 7/13.)
The lateral condyle presents a facet posteriorly for articulation with the head of the fibula at the superior tibio-fibular joint. The condyles are separated behind by the *popliteal notch*.
The *tubercle of the tibia* lies just below the condyles in front. The upper part gives attachment to the patellar tendon, which is the tendon of insertion of the quadriceps extensor muscle. The lower part of the tubercle is subcutaneous, it receives the weight of the trunk in kneeling.

The Shaft is triangular in cross-section. The anterior border is prominent and lies subcutaneously in its middle third, where it forms the *crest of the tibia*. The medial surface is subcutaneous in most of its extent, rendering it a useful area from which to take a *tibial bone graft*. The posterior surface is marked by the *soleal line* which is a strong ridge of bone running downwards and medially.

The Lower Extremity enters into the formation of the ankle joint. It is slightly expanded, and is prolonged downwards on the medial side as the *medial malleolus*. The front of the tibia is smooth, and tendons passing to the foot glide over it.
The lateral surface of the lower extremity articulates with the fibula at the inferior tibio-fibular joint. The tibia articulates with three bones, the femur, fibula, and talus.

FIBULA

The fibula is the lateral bone of the leg. It is a long bone with a shaft and two extremities.

The Upper Extremity forms the *head*, and articulates with the back of the outer condyle of the tibia, but does not enter into the formation of the knee joint.

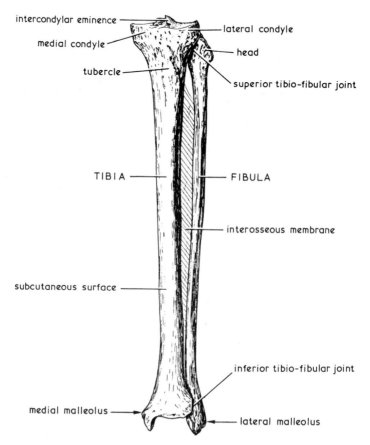

FIG. 6/6.—THE ANTERIOR ASPECT OF THE LEFT TIBIA AND FIBULA

The Shaft is slender and deeply embedded in the muscles of the leg, to which it gives numerous attachments.

The Lower Extremity is prolonged downwards as the *lateral malleolus*.

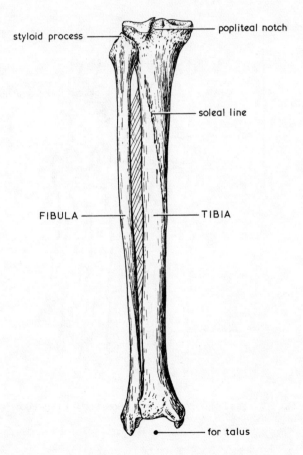

styloid process — popliteal notch

soleal line

FIBULA — TIBIA

for talus

Fig. 6/7.—The Posterior Aspect of the Left Tibia and Fibula

BONES OF THE FOOT

The Tarsal Bones. There are seven bones known collectively
as the *tarsus*. They are short bones, made up of cancellous
bone tissue, with a covering of compact tissue. These bones
support the weight of the body in standing.

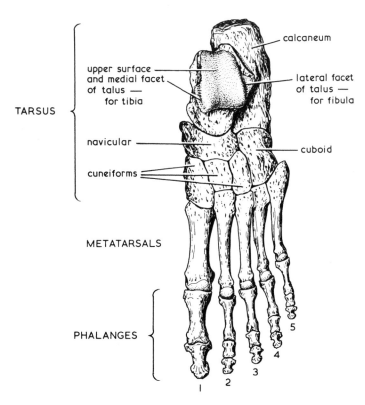

FIG. 6/8.—THE DORSAL ASPECT OF THE BONES OF THE LEFT FOOT

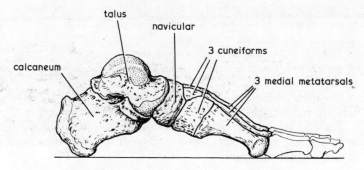

FIG. 6/9.—THE BONES OF THE LEFT FOOT SHOWING THE MEDIAL OR INNER LONGITUDINAL ARCH (*see* Fig. 7/14, page 124)

The Calcaneum or *Os Calcis* is the largest bone of the foot. It lies at the back of the foot forming the heel, transmitting the weight of the body to the ground posteriorly. It gives attachment to the large muscles of the calf through the *tendon of Achilles* or *tendo calcaneus*. Above, it articulates with the talus and in front with the cuboid.

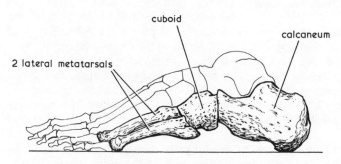

FIG. 6/10.—THE BONES OF THE LEFT FOOT, SHOWING THE LATERAL OR OUTER LONGITUDINAL ARCH

The Talus forms the central and highest point of the foot. It supports the tibia and articulates with the malleoli at each side, and with the calcaneum below.

The Navicular, a boat-shaped bone, lies on the medial aspect of the foot, between the talus at the back and the three cuneiform bones in front.

The three Cuneiform Bones articulate with the navicular posteriorly and with the three inner metatarsal bones anteriorly.

The Cuboid is at the lateral aspect of the foot. It articulates with the calcaneum posteriorly and in front with the two lateral metatarsal bones.

The Metatarsal Bones. There are five metatarsal bones. These are long bones with a shaft and two extremities, the proximal or *tarsal extremity* articulating with the tarsal bones, and the distal or *phalangeal extremity* with the base of the proximal phalanges.

The first metatarsal is thick and short, and the second metatarsal is the longest.

The Phalanges are similar to those of the fingers but much shorter.

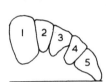

FIG. 6/11.—SECTION THROUGH THE BASES OF THE METATARSAL BONES, SHOWING FORMATION OF POSTERIOR TRANSVERSE ARCH

The Arches of the Foot. Four arches are present in the foot.

The medial or internal longitudinal arch, which is formed from back to front by the calcaneum, the posterior support of the arch; the talus, the summit of the arch; the navicular; three cuneiform bones; and the heads of the three inner metatarsals, forming the anterior support of the arch.

The lateral or *outer longitudinal arch* is formed by the calcaneum, the cuboid and the two outer metatarsal bones.

The transverse arches, of which there are two, *the transverse tarsal arch* formed by the tarsal bones, and the *transverse metatarsal arch,* commonly known as the posterior transverse metatarsal arch, formed by the heads of these bones, the first and fifth forming the piers of the arch (*see* Fig. 6/11). This arch is normally *almost* in contact with the ground in standing, but when the foot is at rest it resumes a more definite shape.

The bones of the arches of the foot are held together by ligaments and supported by muscles. These arches are maintained:

By the *close adaptation* of the bones.

By the *ligaments* of the foot, which are strong.

By *muscular action,* particularly by the muscles attached to the front and back of the tibia.

Clinical Notes

Fractures of the neck of the femur occur as the result of indirect violence, as when a person trips and falls. These fractures are very common in the elderly. Fractures of the shaft may give cause to displacement and over-riding of the fragments, owing to spasm of the large thigh muscles.

The *patella* may be fractured spontaneously by vigorous contraction of the thigh muscles, causing a *transverse fracture.* A *stellate fracture* occurs by falling heavily on to the knee, or by a direct blow on the kneecap.

The *shaft of the tibia and fibula* may be fractured separately or together. The commonest fracture of the fibula is *Pott's fracture,* occurring above the ankle just where the bone gives in the lower third of the shaft; it may be accompanied by dislocation of the ankle joint and also by separation of the medial malleolus of the tibia. The shaft of the fibula is a site of 'stress fracture' in cross-country runners.

Fractures of the bones of the foot, owing to its weight-bearing function, are painful. Any of the tarsal bones, metatarsals and phalanges may be broken. *March fracture* of one of the metatarsals is a 'stress fracture'.

Hallux Valgus is a deviation of the great toe which comes to lie obliquely across the second toe and is frequently associated with a bunion.

The *Arches of the Foot* (*see* pages 109–10). *Flat foot* is due to flattening of the bony arches of the foot; it may follow injuries to foot and ankle, or arise as the result of disturbance of balance which may be traumatic or postural as in deformity of the spine, pelvis or lower limbs. Other causes include *foot strain in walking and standing* (in postmen, policemen, foot soldiers, nurses and others), after illness, or from other causes resulting in muscle weakness.

Depression of the metatarsal heads (the transverse arch) can lead to painful involvement of a digital nerve—a *digital neuroma* (Morton's metatarsalgia).

Chapter 7

THE JOINTS OF THE SKELETON

FIG. 7/1.—TYPICAL
IMMOVABLE JOINT—
SUTURES OF SKULL

A *joint* or *articulation* is the term used to describe the union of any two, or more, bones of the skeleton. The term *arthrology* is applied to the study of joints. There are three main classes: fibrous, cartilaginous and synovial joints.

Or, joints may be classified according to their mobility; immovable, slightly movable and freely movable joints.

Fibrous Joints or *synarthroses*, are immovable or fixed joints in which no movement between the bones is possible:

Sutures or joints of the flat skull bones. In Fig. 7/1 an arrow indicates the *coronal suture,* uniting the frontal and parietal bones; the *sagittal suture* runs from front to back, uniting the two parietal bones, and the *lambdoid* suture unites both parietals with the occipital bone.

Peg and socket joints (gomphosis)—teeth in their sockets and

Syndesmosis, a joint where the articulating surfaces are connected by membrane as in the inferior tibiofibular joint.

Cartilaginous Joints, or *amphiarthroses* are slightly movable joints in which the joint surfaces are separated by some intervening substance and slight movement only is possible: e.g.

The *pubic symphysis*, where a pad of cartilage unites the two pubic

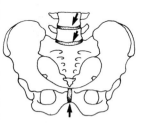

FIG. 7/2.—TYPICAL CAR-
TILAGINOUS JOINTS—
SYMPHYSIS PUBIS AND
INTERVERTEBRAL JOINTS (*see*
FIG. 4/7)

112

bones. The *intervertebral joints* with their intervertebral discs of fibro-cartilage.

The joint between the *manubrium* and the *body of the sternum*.

(A *symphysis* is the term used to describe a partly movable joint, where the bone ends are separated by a pad of fibro-cartilage.)

Temporary or *primary cartilaginous joints* are found between the diaphysis and epiphyses of the long bones before full growth is complete (*see* Fig. 1/17, page 42).

FIG. 7/3.—A TYPICAL SYNOVIAL JOINT

Synovial Joints or *diarthroses* are *freely movable joints* of which there are several varieties, all having similar characteristics (*see* below).

Characters of a Freely Movable Joint

The *ends of bones* which enter into the formation of the joint are covered by *hyaline cartilage*.

Ligaments are required to bind the bones together.

A joint cavity: the cavity is enclosed by a capsule of fibrous tissue which is usually strengthened by ligaments.

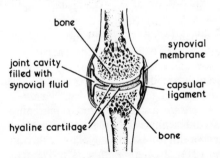

bone

synovial membrane

joint cavity filled with synovial fluid

capsular ligament

hyaline cartilage

bone

Bones are covered by cartilage.

Ligaments bind the bones together.

Synovial membrane lines the joint cavity and secretes fluid to lubricate the joint.

FIG. 7/4.—A SECTION OF A TYPICAL SYNOVIAL JOINT

Varieties of Synovial Joints. There are six varieties. *Gliding joint* or *plane joint,* in which two flat surfaces of bone glide on each other, e.g. the joints of the carpus and the tarsus.

Ball and socket joint, in which one rounded extremity fits into a cavity in another bone, permitting movement in all directions, as in a ball within a socket or cup-shaped cavity, e.g. the hip joint and the shoulder joint.

Hinge joint; in this variety one rounded surface is received into another in such a way that movement is only possible in one plane, as occurs in the movements of a hinge. The best example is the elbow joint.

A *condyloid joint* is similar to a hinge joint, but it is so adapted as to permit movement in two planes, lateral as well as backward and forward, so that flexion and extension and abduction and adduction and slight circumduction are possible, as in the wrist joint, but not rotation.

A *pivot joint* is one in which rotation only is possible, as in the movements of the head, where the ring-like atlas rotates round the peg-shaped process of the axis; another example is seen in the movements of the radius on the ulna in pronation and supination of the forearm (*see* page 118).

A *saddle joint,* or a joint of reciprocal reception, e.g. the joint between the trapezium and the first metacarpal bone of the thumb, permits great freedom of movement, enabling the thumb to be opposed to the fingers.

Movements. The movements taking place at the skeletal joints may be divided into three principal groups.

Gliding movements, in which two flat surfaces move on each other as occurs in the movements between the carpal and tarsal bones.

Angular movements, which are described according to the direction in which the movement takes place—e.g. *flexion,* a bending or doubling up; *extension,* a stretching or straightening out—take place round an axis transversely placed. In the case of the ankle joint the terms dorsi-flexion and plantar-flexion are employed (*see* page 127). *Adduction,* that is movement towards the medial aspect of the body, and *abduction,* in a direction away from the medial aspect of the body, take place round an axis running in an antero-posterior direction—from front to back.

Rotation movements are those in which one bone moves

around or within another bone as in the pivot joints, e.g. the rotation of radius on ulna. It also occurs at the shoulder and to a more limited extent at the hip joint.

Circumduction is the term used to describe a combination of rotation and angular movements, a carrying in circles, e.g. the carrying of the arm forward, upward, backward and downward; including flexion, abduction, extension and adduction, and some rotation.

Limitation of joint movement is due in many instances to the shape of the articulating surfaces, for example extension of the elbow is limited by the olecranon process of the ulna impinging against the humerus. In other instances movement is limited by strong bands of ligaments as in the ilio-femoral ligament on the front of the hip joint which limits extension of the thigh. Flexion of the elbow and of the leg on the thigh are limited by the soft parts coming into contact.

Joints of the Vertebral Arches. *See* page 82.

Temporomandibular Joint. *See* page 72.

JOINTS OF THE UPPER EXTREMITY

The Sterno-Clavicular Joint is a gliding joint formed by the large sternal extremity of the clavicle, articulating with the clavicular facet on the sternum.

The Acromio-Clavicular Joint is formed by the outer end of the clavicle articulating with the acromion process of the scapula.

Shoulder girdle movements. The slight gliding movements possible between the clavicle and scapula, and the play of the scapula on the chest wall are of interest only in so far as they enhance the freedom of movement of the humerus at the shoulder joint.

The Shoulder Joint or the *humero-scapular joint* is a synovial joint of the ball-and-socket variety. The head of the humerus forming one-third of a sphere articulates within the glenoid

cavity of the scapula. The cavity is deepened by the attachment of a fibro-cartilaginous rim, the *glenoidal labrum*. The
bones are united together by ligaments which form a very
loose capsule.

The degree and the limitation of movement here is largely
dependent on the surrounding muscles, and the pressure of
the atmosphere in retaining bones in position whilst looseness of the capsular ligament allows free joint movement in

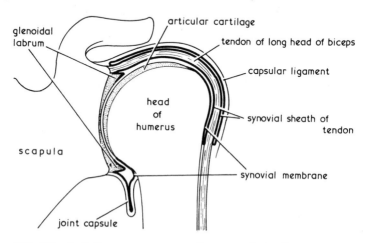

FIG. 7/5.—A SECTION THROUGH THE SHOULDER JOINT, INDICATING ITS
COMPONENT PARTS (semi-diagrammatic)

all directions, abduction, adduction, flexion, extension, medial
and lateral rotation, and circumduction. (*See* Clinical Note,
page 120.)

The Elbow Joint is a hinge joint, between the trochlear surface
on the lower extremity of the humerus, and the trochlear notch
of the ulna. This forms the principal part of the joint, the
humero-ulnar joint. The head of the radius articulates with the
capitulum of the humerus, forming the *humero-radial joint*,
and these four articulating surfaces lie within the joint capsule, the radius being carried backwards and forwards with
the ulna, in the movements of the joint.

The *movements* taking place at the elbow joint are flexion and extension. (*See* Clinical Note, page 121.)

The *carrying angle of the elbow* when the elbow is extended and the forearm and hand supinated is about 170 degrees with the upper arm. This is due to the obliquity of the articulating surfaces between the humerus and ulna. The advantage obtained by this carrying angle is that articles are carried clear of the body.

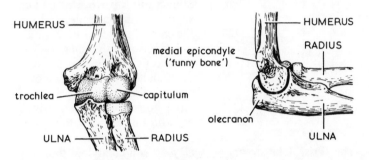

FIG. 7/6.—THE ANTERIOR AND LATERAL ASPECTS OF THE BONES FORMING THE LEFT ELBOW JOINT

The Radio-Ulnar Joints. There are two movable joints between the radius and the ulna, the *superior and inferior radio-ulnar* joints. The interosseous membrane forms a third joint—the *middle radio-ulnar joint* (*see* Fig. 7/7). This membrane also separates the muscles on the front from those on the back of the forearm.

The *movements* of the radius on the ulna are free. As the head of the radius rotates within the annular ligament at the superior radio-ulnar joint, the lower end of the radius rotates on the head of the ulna at the inferior radio-ulnar joint, carrying the hand with it in the movements of *pronation* and *supination* of the forearm.

Pronation is rotation of the radius on the ulna until the hand lies palm downwards. This movement is performed by muscles

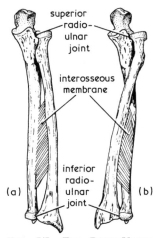

FIG. 7/7.—THE RADIO-ULNAR
JOINTS OF THE RIGHT FOREARM
(a) Supine. (b) Prone.

called pronators which lie on the front of the forearm between the radius and the ulna.

Supination is the opposite movement. Beginning with the forearm prone, it is rotated from within outwards until the radius and ulna lie parallel and the hand lies palm facing upwards. Supination is performed by two supinator muscles situated on the back of the forearm, between the radius and the ulna, and by the biceps muscle inserted into the radial tuberosity. This movement is essential in driving a screw home with a screw-driver, or in turning a door knob.

The Wrist Joint or *radio-carpal joint* is a condyloid joint between the lower end of the radius and the articular disc below the head of the ulna, which together form a concave surface for

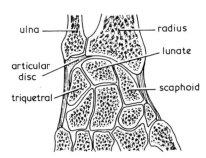

FIG. 7/8.—CORONAL SECTION THROUGH THE WRIST JOINT. THE STRUCTURES FORMING THE WRIST JOINT ARE LABELLED

(*See also* Clinical Note, page 121.)

the reception of the upper aspects of the scaphoid (navicular), lunate, and triquetral bones (*see* Fig. 7/8). The movements of flexion, extension, abduction, and adduction take place at this joint.

JOINTS OF HAND AND FINGERS

The Carpal Joints. The articulating surfaces between the carpal bones are flat and smooth. These flat surfaces move easily on each other, forming *gliding joints* between the different bones. The carpal bones are placed closely together, so that only limited gliding movements are possible, but a fairly considerable amount of movement occurs when all the bones move together.

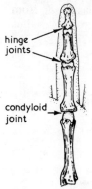

hinge joints

condyloid joint

FIG. 7/9. — THE CARPO-META-CARPAL AND INTERPHALAN-GEAL JOINTS

The Carpo-metacarpal Joints are *gliding joints* formed between the distal aspect of the lower row of carpal bones and the superior articulating surfaces on each of the five metacarpal bones. The carpometacarpal joint of the thumb—saddle-joint—is formed between the base of the first metacarpal and the trapezium. *Inter-metacarpal joints* are formed between the bases of the metacarpal bones; the lateral articulating surfaces form plane or gliding joints between these bones.

The Metacarpo-phalangeal Joints are joints of the condyloid type. The heads of the five metacarpal bones are received into articulating surfaces on the bases of the proximal phalanges.

The *movements* of flexion, extension, abduction, and adduction take place at these joints.

The Interphalangeal Joints are hinge joints. These joints are formed by the heads of the proximal phalanges being received into articulating surfaces on the bases of the distal phalanges.

The *movements* are flexion and extension.

Clinical Notes

In the treatment of an injury or lesion of any joint it is most important to obtain full movement, therefore a note on the movements possible at each joint mentioned is included as of value in promoting rehabilitation.

The **Shoulder Joint,** owing to its shallow articulating cavity, the large head of the humerus and the laxity of the capsular ligament, renders the shoulder more liable to dislocation than any other joint in the body.

(A *dislocation is complete separation of joint surfaces* owing to tearing of the capsule, whereas a *subluxation* is an incomplete separation due to stretching of the capsule.) Dislocation may complicate fractures of the upper extremity of the humerus.

The **Sterno-clavicular Joint** may have a forward or backward dislocation as the result of a heavy fall on the shoulder, e.g. in horse riding.

The **Acromio-clavicular Joints** are more often the site of subluxation than dislocation.

Shoulder Joint Movements. Owing to the looseness of the capsular ligaments and the shallow articulating surfaces, considerable movement is possible. Moreover, movements of the shoulder joint are enhanced by the gliding movements of the scapula on the chest wall.

Abduction is performed by supraspinatus and deltoid, but this is limited to 90 degrees. further elevation to 180 degrees is brought about by rotation of the scapula on the chest wall, principally by the trapezius (*see* Figs. 5/2, 8/4 and 8/10).

Adduction by the weight of the arm and the contraction of muscles in front and behind, principally pectoralis major and latissimus dorsi (*see* Fig. 8/4).

Flexion, carrying the arm forwards and across the chest by pectoralis major and the anterior fibres of deltoid (*see* Fig. 8/4).

Extension by muscles attached to the scapula, teres major, latissimus dorsi and the posterior fibres of deltoid.

Internal and external rotation is also possible and *circumduction* is carrying the arm in circles, up, out, back and down, when most of the muscles attached to the shoulder are brought into action.

It is always important when treating injuries to the shoulder to see that movement to right angles is first performed and then further abduction (*see* note above), so that a patient can do his own hair and wash the back of his neck. It is also essential to obtain this movement after radical mastectomy.

Dislocation of the head of the radius, with displacement forwards, occurs in young people falling on a supinated extended forearm. 'Pulled elbow' is a term used to describe subluxation of the radial head in a young child due to a sudden jerk on his arm.

Elbow Joint. Fractures of the bones forming the elbow joint are often complicated by dislocation. A backward dislocation of the joint may be accompanied by fracture of the coronoid process.

Muscles moving the elbow

Flexion —biceps (*see* Fig. 8/1)
         brachialis
         flexor muscles of forearm } (*see* Fig. 8/11)
Pronation —Pronators and flexor radialis (*see* Fig. 8/11)
Extension —triceps (*see* Figs. 8/1 and 8/12)
         anconeus
Supination—biceps
         supinator (brachio-radialis) (*see* Fig. 8/11)
         extensors of thumb (*see* Fig. 8/12)

Biceps is a flexor muscle of the elbow and, owing to its insertion into the tuberosity of the radius, it rotates the forearm into the position of supination also at the same time. A note on *pronation* and *supination* of the forearm will be found on pages 117–18.

Joints of the Wrist and Hand. The wrist joint may be sprained or strained, requiring support for a time; otherwise there is a tendency to drop things. One or more *carpal bones*, e.g. the lunate, may be dislocated by falling on the hand; the scaphoid or *navicular* may be broken by falling on the palm of the hand.

A fracture of the base of the first metacarpal with subluxation of a part is called a Bennett's fracture.

Muscles moving the wrist
Flexion—the long muscles crossing the front of the wrist (*see* Fig. 8/11)
Extension—all those crossing the back of the joint (*see* Fig. 8/12)
Adduction—the carpal flexor and extensor on the ulnar side of the wrist
Abduction—the carpal flexor and extensors on the radial side (*see* Figs. 8/11 and 8/12)

Muscles acting on the hand are the long flexors and long extensors of the fingers and the small (intrinsic) muscles attached to the carpal bones. Place your hand flat on a table, fingers together and observe that abduction and adduction of the fingers are performed by these intrinsic muscles, assisted in the case of the index and little finger by a special abduction muscle.

The *thumb* is capable of flexion, extension, abduction and adduction and of opposing it to the fingers, as in grasping objects.

JOINTS OF THE LOWER EXTREMITY

The Hip Joint is a synovial joint of the ball-and-socket variety. The head of the femur is received into the *acetabulum* of the innominate bone. The acetabulum is deepened by the attachment of the *acetabular labrum* to its circumference. This ligament is in the nature of a rim of fibro-cartilage which deepens and increases the adaptability of the surface formed by the acetabulum for the reception of the head of the femur.

The *capsular ligament* of the hip joint is thick and strong and limits the movement of the joint in all directions. The ligament

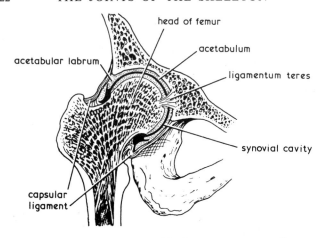

FIG. 7/10.—A SECTION THROUGH THE HIP JOINT, INDICATING ITS COMPONENT PARTS

is also specially strengthened by bands of fibres in several parts. One of the most important of these bands lies in front of the joint, the *ilio-femoral ligament*. This ligament limits extension at the joint, and so helps to maintain the erect position in standing.

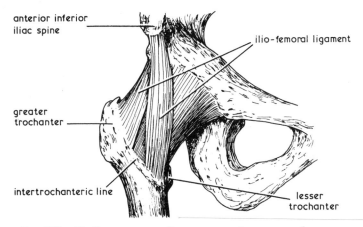

FIG. 7/11.—HIP JOINT, SHOWING POSITION OF THE ILIO-FEMORAL LIGAMENT

The *movements* occurring at the hip joint are flexion, extension, abduction, adduction, medial and lateral rotation. A combination of all these movements is called circumduction. (*See* Clinical Notes, page 125.)

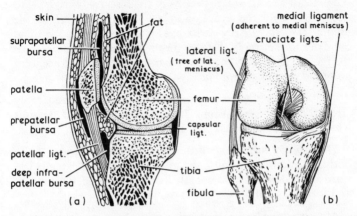

FIG. 7/12.—(a) A Section through the Knee Joint to One Side of the Midline and Semi-diagrammatic. (b) Front View of Flexed Knee showing Chief Ligaments

The Knee Joint is a modified hinge joint formed by the condyles of the femur articulating with the superior surfaces of the condyles of the tibia. The patella lies on the smooth patellar surface of the femur over which it glides during the movements

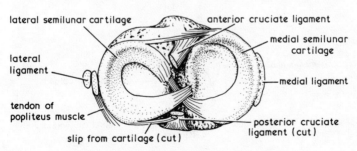

FIG. 7/13.—The Upper Aspect of the Left Tibia

Showing some of the Interarticular Structures.

of the joint. It lies in front of the main articular parts of the joint, but does *not* enter into the formation of the knee joint.

Interarticular structures. Several important structures lie within the knee joint. *The semilunar cartilages* are placed on the plateau-like articular surfaces of the tibia (*see* Fig. 7/13) to deepen these for the reception of the condylar surfaces of the femur. *The cruciate ligaments* pass from the top of the tibial condyles to the rough surfaces on the intercondyloid notch of the femur. These serve to limit the movement of the knee joint and bind the bones more firmly together.

The capsular ligament is extensive and is considerably strengthened by expansions from the muscles and tendons which surround and pass over the joint.

The synovial membrane of the knee joint is the largest in the body. In addition to lining the joint structures, it extends upwards and downwards beneath the ligaments of the patella, and forms several bursae about the joint.

Movements, Flexion, Extension and *slight medial rotation.* (*See* Clinical Note, page 126.)

The Tibio-Fibular Joints. These joints are formed between the upper and lower extremities of the two bones of the leg, the shafts of the bones being united by an interosseous ligament which forms a third joint between these bones as in the forearm (*see* Figs. 6/6, 6/7, pages 106, 107).

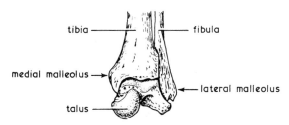

FIG. 7/14.—THE ANTERIOR ASPECT OF THE BONES FORMING THE ANKLE JOINT (RIGHT)

The Ankle Joint is a hinge joint formed between the lower extremity of the tibia and its medial malleolus, and the lateral malleolus of the fibula which together form a socket to receive the body of the talus. The *capsule* of the joint is strengthened by additional important ligaments. The *deltoid ligament* on the medial aspect passes from the medial malleolus to the adjoining tarsal bones and is often badly torn in *severe sprains of the ankle.*

The *movements* of the ankle joint are flexion and extension, or, as more usually expressed, *dorsi-flexion* and *plantar-flexion.* (*See* Clinical Note, page 127.)

JOINTS OF THE FOOT

The joints between the tarsal bones are gliding joints. The bones are united by dorsal, plantar, and interosseous ligaments.

The interosseous ligament placed between the undersurface of the talus and the upper surface of the calcaneum is thick and strong, and grooves the joint surfaces of both these bones.

Movements. A little rocking occurs at the *talo-calcaneal joint*, which is similar to adduction and abduction. The joints between the head of the talus and the navicular, and between the calcaneum and the cuboid, are called the *medio-tarsal joints* or *sub-taloid joints* (*see* Figs. 6/9 and 6/10, page 109.) It is at these joints that the movements of *inversion* and *eversion* take place.

In inversion the inner border of the foot is raised and the sole is directed inwards. In eversion the outer border of the foot is raised and the sole tends to be directed outwards. In conjunction with these movements, slight adduction and abduction occurs at the talo-calcaneal joint. (*See also* Clinical Note, page 127.)

The *tarso-metatarsal*, the *metatarso-phalangeal*, and the *interphalangeal joints* are similar to those described in the hand (*see* page 119).

Clinical Notes

The **Hip Joint** may be dislocated in any direction, most often backwards and medially where the capsule is weakest but, as a rule, the position, extent and complication of a dislocation are determined by the position of the thigh when the impact strikes. For instance, when sitting in a car or train with the knee flexed, any severe impact on the knee from the front of the car or opposite seat in a train may cause the hip to dislocate backwards.

The ilio-femoral ligament crossing the front of the hip joint (*see* Fig. 7/11) is very strong, so that forward dislocation of the hip joint is rare.

Muscles acting on the hip joint. This joint is surrounded by strong, thick muscles, which have to be divided in any operation on the joint.

Flexion—ilio-psoas (*see* Fig. 8/9, page 136) and rectus femoris (*see* Fig. 8/13)

Extension—gluteus maximus and the hamstring (*see* Fig. 8/15, page 142)

Adduction—the group of adductors at the inner aspect of the thigh (*see* Fig. 8/13, page 140)

Abduction—mainly by gluteus medius and minimus

Lateral rotation—gluteus maximus (*see* Fig. 8/15)

Medial rotation—ilio-psoas (*see* Fig. 8/9)

Congenital dislocation of the Hip is commoner than in any other joint. This produces an ungainly gait when the child begins to walk, with disablement in later life. Doctors, midwives and nurses should be able to apply the several tests (e.g. Ortolani's test) soon after birth, so that treatment may be begun without delay.

Knee Joint. Although the articulating surfaces are not well adapted, the knee joint is surrounded by strong ligaments and protected by powerful muscles (its most important factor) which renders it one of the strongest and most stable joints in the body, and one which is rarely subject to traumatic dislocation.

'*Slipped cartilage*'. One of the *semilunar cartilages* (*see* Fig. 7/13, page 123) may be torn, detached and displaced. The accident occurs by a twist of the leg when the knee is flexed, accompanied by pain and often locking of the joint in flexion, because a portion of the torn cartilage becomes lodged between the condyles, preventing extension. Reduction and exercises may help, but *meniscectomy*, removal of the displaced cartilage, which is usually the medial one (the semilunar) is generally necessary to effect a cure.

Muscles acting on the Knee Joint. This joint depends for its stability on its surrounding muscles, particularly quadriceps femoris, which should be kept well developed. The principal muscles acting on the knee are:

Extension—quadriceps femoris (*see* Fig. 8/13)

Flexion—hamstring, gastrocnemius (*see* Fig. 8/15)

Medial rotation—popliteus, a muscle deeply placed at the back of the tibia

Acute Synovitis may result from trauma and, as the synovial membrane is extensive, the accompanying swelling may rise 2·5–5 cm (one or two inches) on each side above the patella.

Bursitis, enlargement and inflammation in one of the bursae about the knee joint, may occur—the bursa between the patella and the skin being most often affected in those who kneel ('Housemaid's Knee').

In disease of the knee joint the hamstring muscles may become contracted, resulting in a flexional deformity.

Ankle Joint. The ligaments of the ankle may be sprained and torn when, for example, by slipping off a kerb, or getting the foot into a rabbit hole, the joint is forcibly inverted or everted (twisted inwards or outwards). There is pain and rapid swelling, which may be limited by an immediate application of cold water firmly bandaged on as a first-aid measure, but a serious sprain should always be X-rayed as one of the malleoli or one of the tarsal bones may have been fractured (*see also* the note on Pott's fracture, page 111).

Muscles acting on the Ankle Joint

Dorsi-flexion—tibialis anterior, assisted by the long extensors of the toes. Flex your ankle and observe contraction of these tendons.

Plantar-flexion— gastrocnemius, assisted by tibialis posterior and the long flexors of the toes.

Muscles acting on the Foot (*see* note on movements above). The muscles principally concerned in maintaining the arches of the foot are *peroneus longus* which passes beneath the sole, *tibialis anterior* from the front, and *tibialis posterior* from the back of the leg forming a double sling or stirrup to support the arches of the foot. The tendons of these three muscles can be seen in relation to the bones of the foot in Figs. 2/14–2/17, pages 61–62.

Arthritis is inflammation of a joint or joints which may attack persons of any age but most frequently those of the middle and elderly age groups.

Rheumatoid arthritis is a polyarthritis, bilateral and symmetrical in distribution, affecting first the small joints of the hands and fingers, osteophytes (Herberden's nodes), forming at the interphalangeal joints.

Treatment. When the joints are painful and inflamed rest is indicated; it is important to arrest the spread of the disease by giving steroids. The joints should be kept as mobile as possible.

Osteo-arthritis is a progressive disease of advancing age; it generally begins as a mono-arthritis, one large joint—hip or shoulder—being affected, but may spread to the knees and other joints. Degenerative changes take place in the cartilages of the joints with lipping at the margins resulting in pain, stiffness and limitation of movement.

Treatment. In this type of arthritis the aim also is to limit the spread of the disease as early as possible, generally by steroids, either by mouth or by injection into the affected joint. It is important to maintain movement by physiotherapy; analgesics are useful in relieving pain. These patients should have their weight controlled by reasonable dieting as it is essential they should not be overweight.

Chapter 8

THE MUSCLES OF THE SKELETON

The muscles of the skeleton form part of one of the four groups of elementary tissues (*see* page 26). *Myology* is the term used to describe the study of muscles.

Muscles are attached to bone, cartilage, ligaments, and to the skin. Those placed immediately beneath the skin are flat. The muscles which surround the trunk are broad and flat, and those of the limbs are long.

The skeletal muscles are sometimes named according to their *shape*, as the Deltoid; according to the *direction* of their fibres, as the Rectus Abdominis; according to the *position* of the muscle, as Pectoralis Major, and according to their *function*, as Flexors, Extensors, etc.

Skeletal muscles are usually attached to two definite points, the more fixed point being named the *origin* and the more movable part the *insertion*. The origin is considered to be the point from which the muscle arises, and the insertion the point to which the muscle passes, the latter point being the structure providing attachment which is to be moved by that particular muscle. With the exception of a very few, each muscle can act on either point; thus the origin and insertion are said to become reversed. For example: Biceps arises from the scapula and passes down the arm to be inserted into the radius; thus the scapula being the more fixed point, the radius is the point moved by the biceps; but if a horizontal bar is grasped by the hands and the body raised to the arms, biceps will contract to assist in this movement, and will then act with the origin and insertion reversed. In this case the radius becomes the more fixed point and the scapula the point to be moved.

The skeletal muscles do not act individually but in groups to perform movements of the different parts of the skeleton. Each group opposes another and is called its *antagonist*. Flexors are antagonists of extensors, abductors of adductors. Certain groups act in stabilizing parts of a limb during movements of other parts; these are called *fixation muscles*. Others partially steady one joint whilst another is moved, as the flexors of the wrist steady it when the fingers are extended. These are called *synergists*.

Tendons e.g. tendon of Achilles, Fig. 8/15, page 142, bind muscle to bone; these are white, glistening, inelastic fibrous bands. *Aponeuroses* are flattened sheets or bands of fibrous tissue, serving as investment for groups of muscles and sometimes connecting a muscle with the part it moves.

Fascia is a mixture of fibrous and areolar tissue found wrapping up, and binding down, the soft structures of the body. *Superficial fascia* lies beneath the skin. It contains fat. *Deep fascia* is dense and more fibrous than superficial fascia. It forms sheaths for muscles and partitions which separate different groups of muscles. In certain parts, as in the palm of the hand, this fascia is very thick and strong.

Palmar fascia. A specially thickened portion of the deep fascia, spread out over the palm of the hand and binding down the deep structures beneath. (*See also* Fig. 8/11.)

Plantar fascia is similarly placed bands of fascia, binding down the structures in the sole of the foot.

Retinacula are thickened portions of deep fascia binding down tendons passing over the wrist and ankle to enter the hands and feet. (*See* Figs. 8/11, 8/12 and 8/14.)

In the following pages some of the principal muscles are shown in diagrams. It is useful to be acquainted with the position of these muscles in relation to the surface of the body and if possible with their main action. Therefore the action of some of the larger muscles in the movements of the joints of the appendicular skeleton are included in the clinical notes given on pages 120–21 and 126–27.

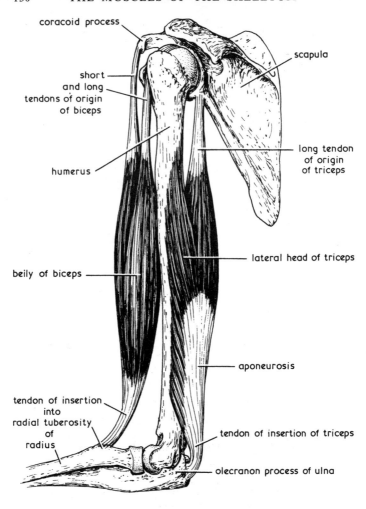

FIG. 8/1.—THE BICEPS AND TRICEPS (LATERAL VIEW OF LEFT ARM). THESE
ARE TYPICAL SKELETAL MUSCLES. NOTE THE TENDONS OF ORIGIN AND
INSERTION, ALSO THE BELLY OF THE MUSCLES

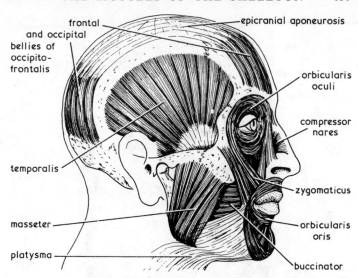

FIG. 8/2.—SOME MUSCLES OF THE HEAD AND FACE

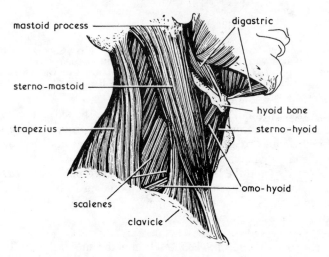

FIG. 8/3.—SOME MUSCLES OF THE NECK (PLATYSMA NOT SHOWN)
(*See also* Fig. 2/1, page 46.)

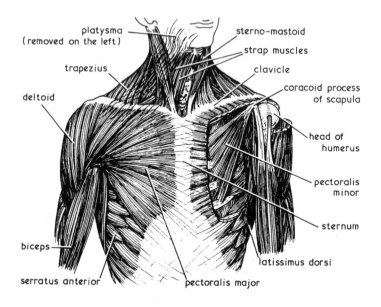

platysma
(removed on the left)

sterno-mastoid

strap muscles

trapezius

clavicle

coracoid process
of scapula

deltoid

head of
humerus

pectoralis
minor

sternum

biceps

latissimus dorsi

serratus anterior

pectoralis major

FIG. 8/4.—THE MUSCLES OF THE ANTERIOR ASPECT OF THE SHOULDER AND
CHEST WHICH WILL BE INVOLVED IN DISEASE OF THE BREAST AND ALSO IN
INJURY TO THE SHOULDER JOINT. PECTORALIS MAJOR AND DELTOID ARE
PARTLY REMOVED ON THE LEFT. FOR THOSE ON THE BACK OF THE SHOULDER,
see Fig. 8/10, page 137

The Diaphragm. The diaphragm is a dome-shaped, musculo-tendinous structure separating the thoracic from the abdominal cavities. It forms the floor of the former and the roof of the latter cavity.

It arises from the lumbar vertebrae by two pillars or crura, from the posterior surface of the xiphoid process, from the inner surfaces of the lower six pairs of ribs and converges to form a central tendinous portion.

Function. In *inspiration* contraction of the muscle flattens the dome of the diaphragm, thus enlarging the vertical diameter of the thoracic cavity. This *descent of the diaphragm* causes air to be drawn into the lungs which expand to fill the enlarged thoracic cavity.

In *expiration* the muscle fibres of the diaphragm relax, and the dome rises, and, as the size of the thoracic cavity is thereby decreased, air is forced out of the lungs.

In addition to being the principal muscle of respiration, the diaphragm also compresses the abdominal viscera, when it descends, and thus assists in the acts of micturition and

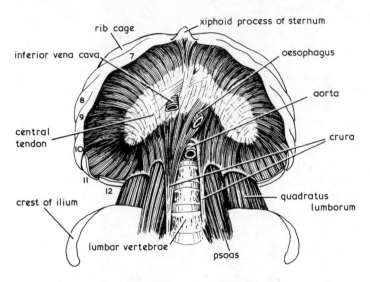

FIG. 8/5.—THE UNDER SURFACE OF THE DIAPHRAGM

defaecation, and in parturition. The height of the diaphragm changes with posture. It is highest when lying down, and lowest when standing or when sitting erect. It is for this reason that patients suffering from *dyspnoea* are more comfortable when sitting up.

There are three openings or *hiatuses* in the diaphragm: the aortic opening for the passage of the aorta and thoracic duct which lies behind the diaphragm between the crura and the vertebral column and does *not* therefore pass through the diaphragm; the *oesophageal* opening through which the

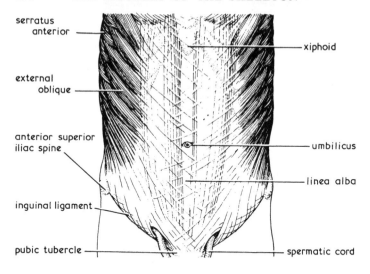

serratus anterior

external oblique

anterior superior iliac spine

inguinal ligament

pubic tubercle

xiphoid

umbilicus

linea alba

spermatic cord

FIG. 8/6.—THE SUPERFICIAL LAYER OF ABDOMINAL MUSCLES

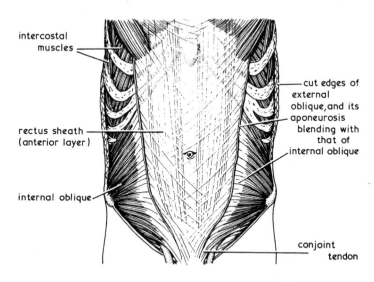

intercostal muscles

rectus sheath (anterior layer)

internal oblique

cut edges of external oblique, and its aponeurosis blending with that of internal oblique

conjoint tendon

FIG. 8/7.—MIDDLE LAYER OF ABDOMINAL MUSCLES

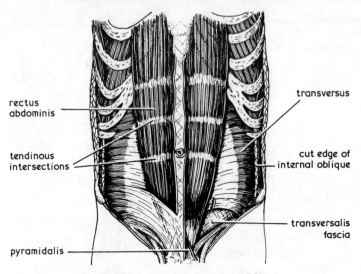

rectus
abdominis

transversus

tendinous
intersections

cut edge of
internal oblique

transversalis
fascia

pyramidalis

FIG. 8/8.—DEEP LAYER OF ABDOMINAL MUSCLES

oesophagus and the vagus nerves pass, and the *caval opening* for the passage of the inferior vena cava.

Relations of the Diaphragm. Above, the apex of the heart and the pericardium, the bases of the lungs and the pleurae.

Below, the liver, stomach, spleen, both suprarenal glands, and both kidneys.

The oesophagus, inferior vena cava, and the vagus nerves pass *through the diaphragm*, the aorta and thoracic duct pass *behind it* (*see* note above).

Nerve supply—Phrenic and intercostal nerves.

Clinical Notes

The *Linea Alba*, or white line, is a line of tendon along the middle of the abdomen from the xiphoid cartilage to the pubis. It separates the two recti muscles. It is interrupted by the *umbilicus* in the fetus, the orifice of which closes a few days after birth. Below the umbilicus this portion of the line becomes fine and faint but during pregnancy it darkens and is termed the *linea nigra*.

The *Inguinal Ligament* (*Poupart's*) is formed by a thickened portion of the lower border of the external oblique muscle of the abdomen. It passes from the anterior superior iliac spine to the pubic tubercle. Beneath it the

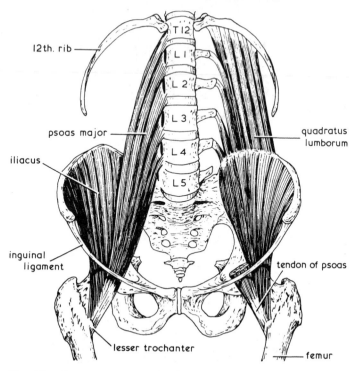

FIG. 8/9.—DIAGRAM SHOWING THE POSITION OF THE PSOAS AND ILIACUS MUSCLES. ON THE RIGHT THE BLENDING OF THE TWO MUSCLES IS SHOWN AS THEY PASS TO THEIR COMMON INSERTION ON THE LESSER TROCHANTER OF THE FEMUR. THESE TWO MUSCLES ARE POWERFUL FLEXORS OF THE HIP JOINT

femoral artery and vein and the femoral nerve pass into the thigh (*see* Fig. 8/18, page 146).

The *Inguinal Canal*, which is about 4 cm (1½ inches) long, is formed in the muscles of the anterior abdominal wall, above the inguinal ligament, directed obliquely medially, downwards and forwards. It contains and transmits the spermatic cord in the male, the round ligament of the uterus in the female; and also some nerves and blood vessels.

The *deep or internal ring* is the point in the fascia of the transversalis muscle where the spermatic cord enters to traverse the inguinal canal, and the *superficial or external ring* is a point in the external oblique abdominal muscle where it emerges to descend to the groin or into the scrotum.

An **inguinal hernia** protrudes through the deep ring, pushing intestine and/or omentum (the contents of the hernia) and the peritoneum (the

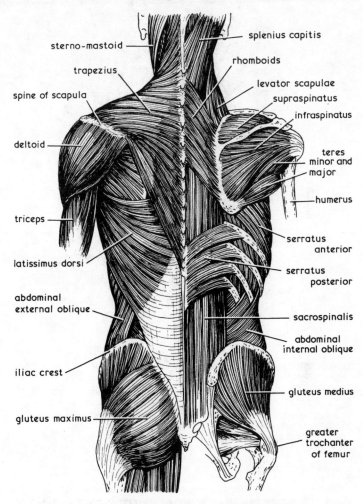

FIG. 8/10.—THE MUSCLES OF THE BACK. ON THE RIGHT THE SUPERFICIAL
MUSCLES HAVE BEEN REMOVED

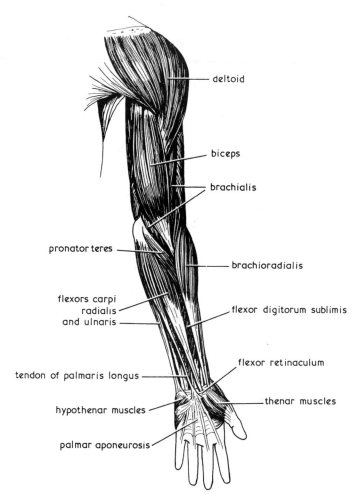

deltoid

biceps

brachialis

pronator teres

brachioradialis

flexors carpi
radialis
and ulnaris

flexor digitorum sublimis

flexor retinaculum

tendon of palmaris longus

thenar muscles

hypothenar muscles

palmar aponeurosis

FIG. 8/11.—SUPERFICIAL MUSCLES ON THE ANTERIOR ASPECT OF THE ARM
AND FOREARM (LEFT)

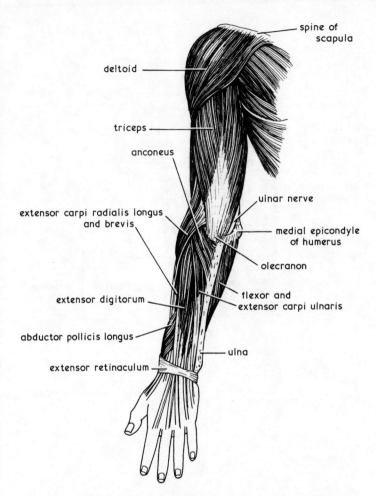

Fig. 8/12.—THE MUSCLES OF THE POSTERIOR ASPECT OF THE ARM AND FOREARM (LEFT), SHOWING ALSO THE EXTENSOR TENDONS TO THE HAND PASSING BENEATH THE RETINACULUM

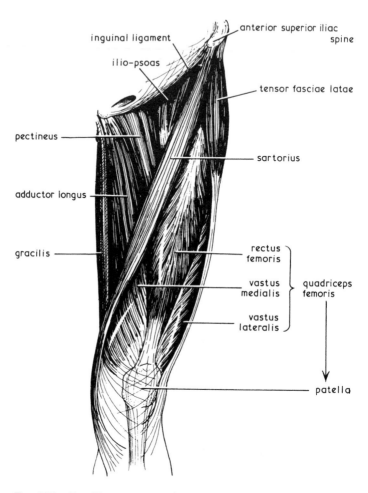

FIG. 8/13.—THE MUSCLES OF THE ANTERIOR ASPECT OF THE THIGH (LEFT)
INDICATING THOSE WHICH CONVERGE FOR ATTACHMENT TO THE PATELLA
(*see also* Fig. 8/14)

Vastus Intermedius, the fourth component of Quadriceps Femoris, lies deep to the others and
therefore is not shown.

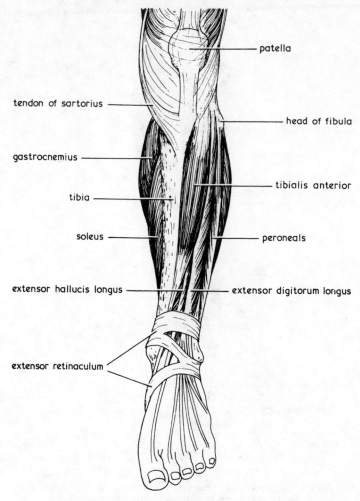

FIG. 8/14. — THE MUSCLES OF THE ANTERIOR TIBIAL OR EXTENSOR ASPECT OF THE LEG (LEFT)

Note the subcutaneous areas of the patella, tibia and malleoli, and the strong retinaculum beneath which pass the extensor tendons to the toes.

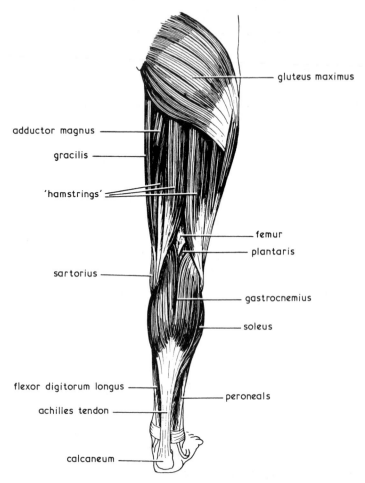

FIG. 8/15.—THE SUPERFICIAL MUSCLES OF THE BACK OF THE THIGH AND
LEG, SHOWING ALSO THE BOUNDARIES OF THE POPLITEAL SPACE

Contracture of the hamstring muscles often complicates knee-joint disease, giving rise to flexion
deformity at the knee joint. Contracture of the tendon of Achilles occurs in drop foot and is
associated with claw foot.

hernial sac) through the canal and superficial ring. The condition may be due to congenital defect or it may be acquired, more often in men than women. In adults a radical herniotomy is necessary to effect a cure.

This hernia may become strangulated on leaving the deep or superficial inguinal ring, when, deprived of its blood supply, it becomes an abdominal surgical emergency.

An **umbilical hernia** may be congenital, though this is rare; it exists normally in the developing fetus but persistence of it is thought probably to be due to a developmental defect.

Acquired umbilical hernia is seen in women of middle age, with a lax abdominal wall, who are overweight. Strangulation may occur.

A **femoral hernia**, commoner in women than in men, usually contains intestine and/or omentum, which is covered by a peritoneal sac, and protrudes through the femoral ring to enter the thigh. Available space is narrow and the structures tight so that there is danger of hernial strangulation. (*See* Fig. 8/18, page 146.)

Posterior Aspect of Lower Abdominal Wall. Muscles entering into the formation of the lower part of the posterior abdominal wall:

Psoas. The upper part of the psoas lies behind the diaphragm in the lower part of the mediastinum: it lies also in relation to the quadratus lumborum (*see* Fig. 8/9). The *lumbar plexus* lies in the substance of this muscle, and the *abdominal aorta*, the *inferior vena cava* and the *receptaculum chyli* and many *lymphatic glands* lie in front of it.

The Iliacus lies on the iliac bone on each side; on the right side the *caecum* lies upon it and on the left side part of the descending colon is in contact with it.

ANATOMICAL SPACES

The Axilla is a pyramidal-shaped space between the arm and the wall of the chest. It is bounded medially by the chest wall and the structures upon it, laterally by the humerus and the muscles attached to it, anteriorly by the pectoral muscles, and posteriorly by the muscles attached to the axillary border of the scapula. *The axilla contains* the axillary artery, axillary vein, the brachial plexus of nerves, and numerous lymphatic vessels

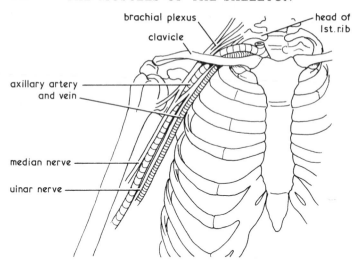

FIG. 8/16.—POSITION OF THE STRUCTURES CONTAINED IN THE AXILLA

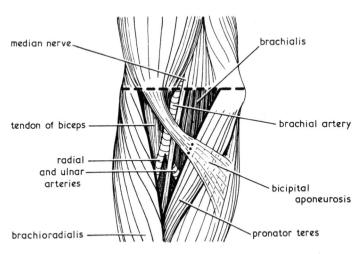

FIG. 8/17.—THE CUBITAL FOSSA (RIGHT)

Note the position of the brachial artery and see page 145. (*See also* note on page 184.)

and glands draining the arm, hand and chest wall (*see* Fig. 8/16 and Fig. 12/2, p. 196). The *axilla* also contains a prolongation of the breast in the female known as the axillary tail. Much of the lymphatic drainage of the breast is to glands in the axilla. Thus a malignant growth in the breast can spread to and enlarge these lymphatic glands. No clinical examination of the breast is complete until the axilla has also been palpated carefully.

The Cubital Fossa is a space at the bend of the elbow. It is bounded above by an imaginary line drawn transversely across the lower end of the anterior surface of the arm, medially by the pronator teres muscle, and laterally by the brachio-radialis. The floor of this cavity is formed by the brachialis muscle. It contains the brachial artery (and is a convenient point at which to listen to the pulsation of this artery when taking the blood pressure), the median nerve, and the tendon of the biceps muscle (*see* Fig. 8/17).

The Ischio-rectal Fossa is a space between the ischium and the rectum. It is filled with connective tissue and fat. *An ischio-rectal abscess* may arise by infection spreading from the rectum as in a case of infected haemorrhoids.

The Femoral (Scarpa's) Triangle is placed immediately below the inguinal (Poupart's) ligament which forms the base of the triangle; it is then bounded laterally by the sartorius muscle and medially by the adductors of the thigh. The floor is formed by the deep muscles of the thigh. It contains the femoral artery, femoral vein, femoral nerve, and lymphatic vessels and glands (*see* Fig. 8/18, page 146).

The Subsartorial (Hunter's) Canal is a passage running along the front and medial aspect of the thigh to reach the back. It extends from Scarpa's triangle to the popliteal space. The femoral artery and deep femoral vein pass through this canal.

The Popliteal Space lies at the back of the knee joint, the posterior surface of which forms the floor of the space. It is a

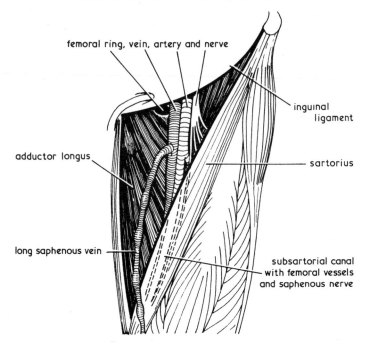

FIG. 8/18.—THE FEMORAL TRIANGLE (LEFT), SHOWING THE STRUCTURES
CONTAINED IN IT

The space to the inner side of the femoral vein constitutes the femoral canal, closed by the
femoral ring, through which a femoral hernia (*see* page 143) passes in to the thigh.

diamond-shaped space bounded above by the medial and
lateral hamstring muscles and below by the medial and lateral
heads of gastrocnemius. It contains the popliteal artery and
vein, the medial and lateral popliteal nerves, and several small
lymphatic glands (*see* Fig. 8/15).

Clinical Notes

Myopathy is a term used to describe any disease or disorder of the skeletal muscles which is thought to be due to some inherent error of muscle metabolism. The *chief symptom* is increasing muscular weakness; sometimes muscle fibres are replaced by fat, giving an enlargement but still accompanied by increasing weakness, as in *pseudohypertrophic dystrophy*.

Myositis is the term used to indicate any inflammation or disease of a skeletal muscle. There are a number of varieties, some acute, others chronic as in *myositis ossificans* where the muscle fibres are first replaced by fibrous tissue and later by bone, as the name implies.

Cramp is an involuntary, painful, localized contraction of muscle, which can often be relieved by stretching the muscle, as cramp in the calf or thigh may be relieved by forcibly extending the limb with the toes dorsiflexed. Cramp occurs in normal people after vigorous exercise, and at night; it also occurs in certain metabolic disorders, as in sodium depletion, in serious water depletion (*see* page 20) and in certain diseases affecting the motor neurones.

The **Fingers** are one of the exposed parts of the body and through constant use are very liable to injury and infection. All abrasions need careful attention. Infection spreading from the pulp spaces (the soft cushions at the end of the fingers and toes) along the synovial tendon sheaths may spread some distance and requires very serious and careful treatment. An acute infection similarly spreading may involve the lymphatic vessels and glands. Moreover the fingers are very liable to contracture after injury (*see* Contracture of muscles below).

Muscles may be injured by bruising, laceration, sprain or rupture. A muscle may be completely torn across and ruptured. A *haematoma* may form in an injured muscle. In *tennis leg* the fibres of a muscle in the calf are ruptured. *Tennis elbow* is a similar condition when there is damage, due to strain, of the common extensor muscle origin from the lateral epicondyle of the humerus, when any movement of these extensor muscles causes pain. A muscle, e.g. the rectus abdominis may be the site of *tumour*.

Contracture of muscles may occur after any injury, particularly after burns unless care is taken in maintaining the affected muscles in their normal acting position by adequate splinting. Contraction may also be due to other causes; an example is seen in *torticollis* when, owing to contracture of the sternomastoid of one side the head is flexed and the face rotated to the opposite side. This may be a congenital condition or it may be due to spasm of the muscle arising as the result of some form of irritation.

Volkmann's ischaemia is a condition caused by interference to the blood supply of a muscle or group of muscles, such as those in the forearm. The muscle fibres undergo fibrosis and become contracted. Too tight application of plaster of paris is the usual cause.

Tendons also may be injured by stretching, tearing or rupture; *tenosynovitis* follows if an injured tendon is infected. A tendon may be divided by being cut across in an accident or street fight, by razor blades for example. Tendons may also be contracted, one classical example being *Dupuytren's* *contracture* when the palmar fascia is drawn up and the ring and little finger are flexed into the palm of the hand and cannot be extended, either actively or passively; unless treated (and surgical measures may be needed), other fingers may be similarly involved as the contracture increases.

The **diaphragm** is also a muscle. It may be injured in an accident involving the chest or abdomen; *it may be paralysed* as happens in accidents to the spine and similarly in poliomyelitis when the phrenic nerves are involved. Some form of artificial respirator is then employed to maintain the movements of respiration.

Certain defects occur in the diaphragm resulting in *congenital herniae*. An *oesophageal hiatus hernia,* owing to muscular weakness, is the commonest type.

The hiatus herniae are acquired also in people of middle age where weakening and widening of the oesophageal opening occurs; in a *sliding hiatus hernia* the lower oesophagus and upper part of the stomach slide up in to the chest when the patient stoops, lifts or stretches. In time *oesophagitis* follows and there is generally some slight but consistent bleeding, which may only be diagnosed when the patient becomes anaemic —an iron-deficiency anaemia, which needs correcting.

Another type is the *rolling hiatus hernia,* when the fundus of the stomach passes up through the opening in front of the oesophagus. Flatulence and epigastric discomfort are characteristic of this type (*see* description of diaphragm, pages 132–35).

Chapter 9

THE CIRCULATORY SYSTEM

The circulatory system consists of the heart, blood vessels and lymphatics.

The *heart*, which is the great pumping organ maintaining the circulation throughout the body,

Arteries carrying blood *from* the heart,

Veins carrying blood *to* the heart,

Capillaries uniting the arteries and the veins and forming the 'capillary lake' where the traffic between nourishment and waste matter proceeds and the interchange of gases takes place in the extracellular or interstitial fluid,

Lymphatics, which collect, filter, and pass back to the blood stream the lymph which has exuded through the minute capillary walls to bathe the tissues, may also be regarded as part of the circulatory system (*see* Chapter 12).

HEART

The heart is a cone-shaped, hollow, muscular organ, having the base above and the apex below. The apex inclines towards the left side. The heart weighs about 300 grams.

Position of the Heart. The heart lies in the thorax, between the lungs and behind the sternum, and directed more to the left than the right side. The exact position may be marked on the body.

A line drawn from the third right costal cartilage 12·5 mm ($\frac{1}{2}$ inch) from the sternum, upwards to the second left costal cartilage 18 mm ($\frac{3}{4}$ inch) from the sternum, marks the position of the *base of the heart* where blood vessels enter and leave (*see* Fig. 9/1).

A point marked on the left side between the fifth and sixth left ribs or in the fifth left intercostal space 9 cm ($3\frac{1}{2}$ inches) from the mid-line, gives the position of the *apex of the heart*, which is the pointed extremity of the ventricles.

149

By uniting these two markings by lines, as shown in the accompanying diagram, the position of the heart may be indicated.

Structure of the Heart. The heart is about the size of a closed fist. The adult heart weighs about 220–260 g (8–9 oz.). It is divided by a septum into two sides, right and left. There is normally no communication between these two sides after birth. Each side of the heart is further subdivided into two chambers,

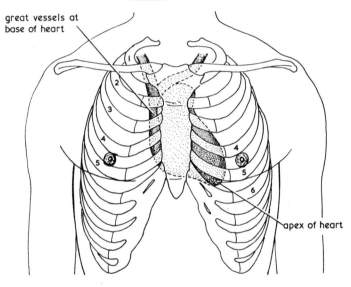

great vessels at base of heart

apex of heart

FIG. 9/1.—THE POSITION OF THE HEART IN RELATION TO THE STERNUM, RIBS AND COSTAL CARTILAGES

an upper chamber called an *atrium,* and a lower chamber, a *ventricle.* There are two atria, right and left, and two ventricles. The atria and ventricles of each side communicate with one another by means of the *atrioventricular openings,* which are guarded by valves; on the right side by the *tricuspid valve* and on the left the *mitral valve.* (The terms atrium and auricle are synonymous.)

The atrioventricular valves permit of the passage of blood in one direction only, i.e. from atrium to ventricle; and they

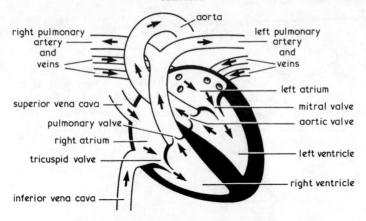

FIG. 9/2.—DIAGRAM OF THE CIRCULATION THROUGH THE HEART
The direction of the blood flow is indicated by arrows.

prevent the blood flowing backwards from ventricle to atrium. The *tricuspid valve* is composed of three flaps or cusps, and the *mitral* of two flaps, which gives it some resemblance to a bishop's mitre, hence the name.

The heart is composed of a specialized cardiac muscle, briefly described on page 33 and is surrounded by a membrane called the *pericardium*. There are two layers: the *visceral pericardium,* a serous membrane, which is closely adherent to the heart, and the *parietal pericardium*, a fibrous layer which is reflected back from the base of the heart, and surrounds it like a loose sac. By this arrangement the heart lies in a double sac of pericardium with serous fluid between the two layers which, by a lubricating action, allows the heart to move freely.

The heart is lined by endothelium; this layer is called the *endocardium*. The valves are simply thickened portions of this membrane. (*See* Clinical Notes, page 160.)

The thickness of the heart wall is composed of a network of heart muscle fibres, and is known as the *myocardium*.

The heart may thus be described as consisting of three layers:
The *Pericardium,* or outer covering,
The *Myocardium,* the middle muscular layer, and
The *Endocardium,* the inner lining.

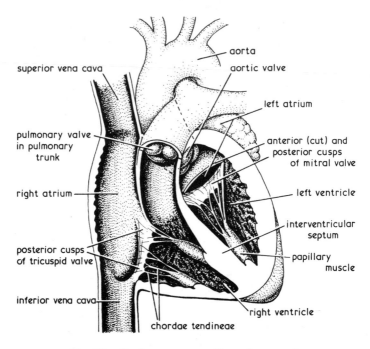

FIG. 9/3.—THE INTERIOR OF THE HEART (SIMPLIFIED)

The muscular walls of the heart vary in thickness: the ventricles have the thickest walls; the walls of the left are thicker than those of the right ventricle, because the force of contraction of the left ventricle is much greater. The walls of the atria are composed of thinner muscle.

The interior of each of the ventricular walls is marked by thickened columns of muscle. Some of these project as papillae, the *papillary muscles*, and to the apices are attached thin tendinous cords, the *chordae tendineae*. These cords have a second attachment to the lower borders of the atrioventricular valves, and this attachment prevents the flaps of the valves from being forced up into the atria, when the ventricles contract (*see* Fig. 9/3).

The Blood Vessels attached to the Heart. The *superior and inferior venae cavae* empty their blood into the right atrium. The opening of the latter is guarded by the semilunar valve of Eustachius. The *pulmonary artery* carries blood away from the right ventricle. The *four pulmonary veins* bring blood from the lungs to the left atrium. The *aorta* carries blood away from the left ventricle.

The openings of the aorta and the pulmonary artery are guarded by the *semilunar valves*. The valve between the left ventricle and the aorta is called the *aortic valve* and prevents blood flowing backwards from the aorta to the left ventricle. The valve between the right ventricle and the pulmonary artery is called the *pulmonary valve* and prevents blood flowing backwards from the pulmonary artery into the right ventricle.

Blood Supply and Nerve Supply of the Heart. The right and left *coronary arteries* are the first to leave the aorta; these then divide into smaller arteries which encircle the heart and supply blood to all parts of the organ. The return blood from the heart is collected mainly by the *coronary sinus* and returned directly into the right atrium.

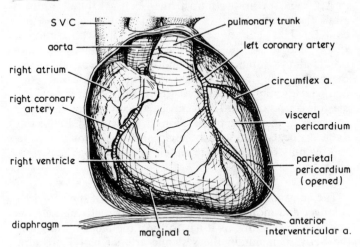

FIG. 9/4.—THE CORONARY ARTERIES AND PERICARDIUM

Nerve supply. Although the action of the heart is rhythmic in character (*see* page 156), its rate of contraction is modified by impulses reaching it from the vagus and sympathetic nerves. Branches from these nerves pass to the sino-atrial node. Control from the sympathetic system *accelerates* the rate of the heart beat, and control by the vagus, which is part of the para-sympathetic or autonomic system (*see* page 363), causes the action of the heart to be slowed or *inhibited.*

Normally the heart is all the time being inhibited by the vagus; but when the vagal tone or 'brake' is removed to meet the needs of the body during exercise or emotional excitement, the *rate of the heart beat increases*; conversely, during physical rest and emotional tranquillity, it decreases. (*See also* Arterial Pulse and Cardiac Output, page 156.)

THE CARDIAC CYCLE

The heart is a pump and the events which occur in the heart during the circulation of the blood are spoken of as the *cardiac cycle*. The heart's action originates in the sino-atrial node (S.A.), then the atria contract, the electrical impulse moves along the bundle of His (*see* Fig. 9/5) and the ventricles then contract. This action is described in two parts, contraction or *systole* and relaxation or *diastole*. Contraction of the atria occurs almost simultaneously and is called the atrial systole; their relaxation, the atrial diastole. Similarly the contraction and relaxation of the ventricles are the ventricular systole and diastole respectively. The ventricular contraction lasts 0·3 seconds; the relaxation phase is longer, 0·5 seconds. In this way the heart beats continuously, night and day, throughout life, and the only rest the cardiac muscle gets is during the periods of ventricular diastole.

The contraction of the atria is short, that of the ventricles is longer and more forcible, and that of the left ventricle the most forcible of any, as it has to force the blood throughout the body to maintain the systemic arterial blood pressure. The right ventricle, although it pumps exactly the same volume of blood, only has to send it round the lungs where the pressure is much less.

The Heart Sounds. Two sounds may be heard during the action of the heart due to the passive closing of the valves.

The first sound is due to the closing of the atrioventricular valves, and the contraction of the ventricles; *the second* to the closing of the aortic and pulmonary valves, after the contraction of the ventricles. The first is long and dull, and the second

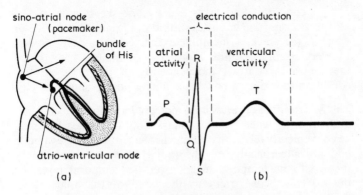

FIG. 9/5.—(a) DIAGRAM OF THE HEART, SHOWING THE ELECTRICAL CONDUCTION SYSTEM (b) NORMAL E.C.G.

short and sharp. Thus the first sounds like 'lubb' and the second like 'dup'. Normally the heart makes no other noise, but if the flow of blood is rapid or if there are deformities in the valves or other chambers there may be extra noises, usually called 'murmurs'.

The Cardiac Impulse or *apex beat* is the impact of the left ventricle against the anterior wall of the chest, occurring during the contraction of the ventricles. This impulse can be felt, and often seen, in the fifth left intercostal space, about 9 cm ($3\frac{1}{2}$ inches) from the middle line of the sternum.

Properties of Heart Muscle. Cardiac muscle has its characteristics (*see* page 33).

Contractility. By contracting the muscle of the heart pumps out of its chambers the blood which enters during diastole.

Conductivity. The contraction is conveyed (conducted) along every individual fibre of the heart muscle with perfect smoothness. This property is very marked in the bundle of His (*see* page 154).

Rhythm. Cardiac muscle possesses the inherent power of automatic rhythmic contraction, independent of its nerve supply (*see* page 34).

In a condition known as *heart block* the bundle of His fails to transmit the impulses started at the sino-atrial node or sinus. If the block is 'incomplete' the ventricles only respond to every second or third impulse. In 'complete' heart block the ventricles contract independently of the atria. In this condition they obey a new 'pace-maker' in the bundle of His.

The Arterial Pulse is a wave of increased pressure which is felt at the arteries when blood is pumped out of the heart. It may be conveniently felt at any point where an artery crosses a bone and lies superficially, as: the radial artery at the front of the wrist, the temporal artery over the temporal bone, or the dorsalis pedis artery at the bend of the ankle. It is *not* the blood pumped by the heart into the aorta that is felt, but the pressure transmitted from the aorta which travels more rapidly than blood.

The pumping rate of the heart varies in health under conditions of living, working, food intake, age and emotion. The pulse rate corresponds with the cardiac cycle (*see* page 154). If the pulse count is 70, the cardiac cycle will occur 70 times a minute.

Normal Pulse Rate Range (*number of beats per minute*)

In the newly born	140	At the age of 5 years	96–100
During the first year	120	At the age of 10 years	80–90
During the second year	110	In the adult	60–80

The Cardiac Output. In a 'resting' person the heart beats about 70 times a minute and pumps 70 ml each time (the stroke volume is 70 ml). The amount of blood pumped each minute is therefore 70×70 ml or about 5 litres.

During exercise the heart rate may be 150 per minute and the stroke volume over 150 ml, making a cardiac output of 20 to 25 litres per minute.

An exactly equal volume of blood is returned to the heart in the veins each minute. If, however, the venous return is not well balanced and the ventricles fail to deal with the cardiac output, heart failure occurs. The large veins near the heart become distended with blood as the venous pressure rises and unless this condition can be dealt with quickly oedema occurs.

The **oedema of heart failure** is partly due to the back-pressure in the veins, which increases the filtration of fluid in the capillaries and partly due to the low cardiac output which reduces the blood flow to the kidney; the kidney then fails to excrete sodium. Retention of sodium causes retention of water (*see* page 21).

THE CIRCULATION OF THE BLOOD

The heart is the chief organ of the circulation of the blood. The course of the blood from the left ventricle through arteries, arterioles and capillaries, returning it to the right atrium by veins is called the greater or *systemic circulation*. The course from the right ventricle, through the lungs, to the left atrium is the lesser or *pulmonary circulation*.

The Systemic Circulation. The blood leaves the *left ventricle* of the heart by the *aorta*, the largest artery in the body. This breaks up into smaller arteries which carry the blood to the different parts of the body. These divide and subdivide until the arterioles are reached. These have very muscular walls which narrow their channels and resist the flow of blood. This has two functions: it maintains the arterial blood pressure and —by varying the size of the channel—it regulates the flow of blood into the capillaries. The capillaries have very thin walls so that exchange can take place between the plasma and the interstitial fluid. These capillaries then unite and form larger vessels called venules which in turn become veins, and carry the blood back to the heart. The veins unite and unite again

until finally <u>two large</u> <u>venous trunks are formed, the</u> <u>*inferior*</u>
<u>*vena cava* which collects the blood from the trunk and lower</u>
extremities and the <u>*superior vena cava* which collects blood</u>
<u>from the head and upper extremities.</u> <u>Both these vessels empty</u>
<u>their contents into the *right atrium* of the heart.</u>

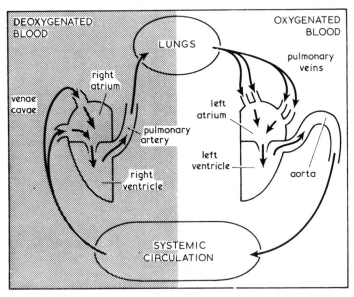

FIG. 9/6.—A DIAGRAM OF THE CIRCULATION

The Heart is shown separated into right and left sides. The arrows indicate the direction of blood
flow (*see* page 157).

Pulmonary circulation. <u>This blood then passes into the *right*</u>
<u>*ventricle* which contracts and pumps it into the *pulmonary*</u>
<u>*artery*.</u> <u>This divides to carry the blood to right and left lungs.</u>
The lungs offer very little resistance to the blood in the vessels
flowing through them. <u>In the lungs each artery breaks up into</u>
<u>numerous smaller arteries, then into arterioles and finally into</u>
<u>*pulmonary capillaries* which surround the alveoli in the lung</u>
tissue where the blood takes up oxygen and gives off carbon
dioxide (for functions of lungs *see* pages 261–63).

<u>The pulmonary capillaries then unite until veins are formed</u>

and the blood is returned to the heart by four *pulmonary veins* which empty into the *left atrium*. This blood then passes into the *left ventricle* which contracts and pumps it into the aorta to begin the *systemic circulation* again (*see* Fig. 9/6).

Pulmonary oedema accompanies failure of the left side of the heart. Tissue fluid collects in the lungs, whose function is impaired. Pulmonary oedema can also occur if a patient who is ill is overhydrated; his lungs become waterlogged and it is possible that he may drown in his own pulmonary oedema.

Portal circulation. Blood from the stomach, intestines, pancreas and spleen is collected by the *portal vein* (*see* Fig. 9/7). In the liver this vein breaks down into a capillary system and,

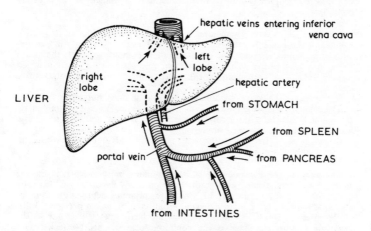

FIG. 9/7.—A DIAGRAM OF THE PORTAL CIRCULATION
The Liver is shown divided into right and left lobes, the Portal Vein dividing to enter both lobes.

uniting with capillaries of the *hepatic artery,* which brings blood from the aorta to the liver, traverses the substance of this organ. This dual blood supply is collected by a system of veins which unite to form the *hepatic vein* conveying the blood to the *inferior vena cava* and thence to the heart.

Portal obstruction may occur when a branch or branches of the portal vein are obstructed, in severe injury to the liver and in some instances in hepatitis. When severe, such an obstruction is complicated by *ascites*, a collection of excess fluid in the peritoneal cavity.

The Coronary Circulation supplies the heart (*see* page 153).

Clinical Notes

Pericarditis is an inflammation of the covering membrane of the heart and of the sac in which the heart lies. These membranes may produce fluid which collects as a pericardial effusion and which embarrasses the heart's action and may require aspiration. After the acute phase, the pericardium may become thickened and tough when constriction of the cardiac action and filling occurs—*constrictive pericarditis*.

This is one cause of heart failure and surgical resection of a large portion of the pericardium may lead to very dramatic improvement in the patient's condition.

Endocarditis. The endocardium is the lining membrane of the inside cavities of the heart and this may become inflamed, particularly in association with rheumatic fever. The condition most commonly affects the mitral valve in young people. The inflammation resolves leaving scarring which usually results in narrowing of the valve orifice producing mitral valve stenosis.

If, however, the valve is damaged so that it does not close fully, *incompetence of the valve* results. In some cases of 'mixed mitral valve' disease both stenosis and incompetence are present at the same time. The aortic valve may be damaged similarly.

Calcification is frequently the end result of valvular endocarditis, in which case the valve may require replacement.

Coronary artery disease. These vessels, like arteries elsewhere in the body, may become narrowed gradually by atherosclerosis or suddenly by a thrombus. The myocardium in either case becomes deprived of some of its blood supply—*myocardial ischaemia*—and pain or *angina pectoris* results. If the artery is totally occluded, part of the heart muscle dies due to the *myocardial infarction*. This is the common form of 'heart attack'—with severe chest pain and failure of the circulation. Special supportive techniques and nursing care can help to salvage a number of these apparently hopeless cases. 'Coronary care' units are now being developed in major centres.

Syncope means sudden loss of consciousness due to cerebral anoxia resulting from cerebral ischaemia. This may be due to a sudden fall in blood pressure (simple faint) or as a result of heart disease itself.

Congestive heart failure is characterized by dyspnoea and the accumulation of oedema fluid in the soft tissues, due to failure of cardiac pumping action. The fluid collects in the most dependent parts of the body, ankles, sacrum or scrotum, according to the patient's position.

Either the right ventricle may fail, producing the clinical picture of high venous pressure, engorged liver with peripheral oedema, or the left ventricle where the back-pressure builds up in the lungs with the development of pulmonary oedema (congestion). (*See also* The oedema of heart failure, page 157. Note on pulmonary oedema, page 159.)

At a late stage this increase in pulmonary congestion may cause even the right ventricle to fail as a secondary phenomenon and combined left and right heart failure ensues. Mitral, aortic, pulmonary and tricuspid valve lesions are the commonest causes of congestive heart lesions.

Cardiac arrest is a very serious emergency demanding immediate relief, as if the brain is deprived of blood for more than 3 to 4 minutes, irreversible damage may follow. *The airway must be kept clear* and **exhaled air respiration** given by mouth to mouth or mouth to nose. The heart cannot beat unless and until the lungs are ventilated.

External cardiac massage can be started simultaneously. With the patient lying on his back on a firm surface, the heel of the hand is placed over the lower part of the sternum, depressing it 50 to 60 times a minute quite regularly. The sternum can be depressed at this point 2·5 to 4 cm (1 to 1½ inches) in the adult. Every nurse should be familiar with this treatment of cardiac arrest and be able to get the treatment going until medical aid arrives. (*See* note on *heart block*, page 156, and *portal obstruction*, page 160.)

Cardiac surgery. Operations inside the heart itself may be performed as closed or 'blind' operations and these are done largely by feel, using the finger and instruments.

With the modern development of the heart lung machine, it is possible to take over a patient's cardiac and pulmonary function entirely for several hours by using *extracorporeal circulation* through the 'pump'. This enables the surgeon to arrest the heart if necessary and to open any of the chambers, to visualize the exact pathological condition and to deal with it. Thus damaged valves may now be replaced and congenital defects corrected.

Heart valves are at present replaced by either a 'prosthetic man-made' mechanical valve, or by a homograft or heterograft valve taken from a human being or animal soon after death.

Many children born with congenital defects, such as atrial or ventricular septial defects, blue-babies with Tetralogy of Fallot and more complicated cardiac abnormalities, may be completely cured following cardiac surgery. The transformation of a stunted, blue, incapacitated child to a pink, active developing adolescent, is one of the miracles of modern medicine.

Valve replacement usually involves adults, many in the fourth and fifth decades of life, who have developed 'acquired heart disease' following rheumatic fever or other specific illnesses.

The **selection of patients** for any operation on the heart and its great vessels demands care. A number of essential tests are first carried out, including electrocardiography, cardiac catheterization, angiocardiography. The surgeon will explain to a patient (or if a child, to parents) what he can do or hope to accomplish, and he will indicate the extent of convalescence which will follow. It may be necessary to support the patient postoperatively with intermittent positive pressure respiration, tracheostomy, drugs which stimulate the heart's action and electrical pacing of the heart with a pacemaker.

Nurses have their place in the team of medical and surgical experts, anaesthetists, biochemists, pathologists, physiotherapists, engineers and others who, chosen for their nursing skill and experience, are beside the patient in the proportion of 3 to 5 covering the 24 hours in an intensive care unit.

The importance of understanding the helplessness of this extremely ill patient cannot be over-estimated. Trained observers are essential, geared with knowledge to decide when specialized aid should be summoned. Bringing comfort and consolation to relatives at all times, and to the patient when he has made enough progress to be interested, is an essential part of the nurse's duties.

Chapter 10

THE BLOOD

Blood is a fluid tissue composed of two parts. The intercellular substance is a fluid called plasma, in which float formed elements—the blood cells or corpuscles. The *total volume* of blood forms about one-twelfth of the weight of the body or about 5 litres. About 55 per cent, a little over half the volume, is fluid, the remaining 45 per cent of the volume being made up of the blood cells. This figure is described as the *haematocrit* or packed cell volume, ranging from 40 to 47.

The volume of blood is constant in health, being regulated to a great extent by the osmotic pressure in the vessels and in the tissues (*see* page 20.)

Composition of Blood. *Blood serum* or *plasma* is made up as follows:

Water	91·0 per cent	
Protein	8·0 ,, ,,	(Albumin, globulin, prothrombin and fibrinogen)
Salts	0·9 ,, ,,	(Sodium chloride, sodium bicarbonate, salts of calcium, phosphorus, magnesium, and iron, etc.)

The balance is made up of traces of a number of organic materials: glucose, fats, urea, uric acid, creatinin, cholesterol, and amino-acids.

The plasma also carries:
Gases—oxygen and carbon dioxide,
Internal secretions,
Enzymes, and
Antigens.
Blood cells. There are three varieties:
Erythrocytes or red cells,
Leucocytes or white cells, and
Thrombocytes or platelets.

The Red Cells or Erythrocytes are small circular bi-concave discs, so-called because they are concave on both sides, so that when looked at from the side they appear like two crescents placed back to back. There are about 5,000,000 red cells in each cubic millimetre of blood (*see* page 169). They are a pale buff colour when seen singly, but in masses appear red and give the colour to the blood. In structure they consist of an outer envelope or stroma which encloses a mass of *haemoglobin.*

The red blood cells need *protein* for their structure derived from the amino-acids; they also need *iron*, so that a balanced diet containing some iron is necessary for their replacement. Women require more iron as some is lost in the menstrual flow; in pregnancy the requirements are greater to supply iron for the developing fetus and for the milk in lactation.

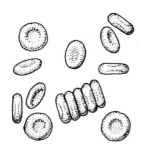

FIG. 10/1.—RED BLOOD CELLS; SOME ARE IN ROULEAU FORMATION, SHOWING THAT THESE CELLS ARE BI-CONCAVE DISCS

The red cells originate in bone marrow, especially in that of the short, flat, and irregular bones, in the cancellous tissue at the ends of the long bones and in the marrow in the shafts of the ribs and in the sternum.

In process of development in the bone marrow the red cells pass through several stages: at first they are large, and contain a nucleus but no haemoglobin; they are next charged with haemoglobin and finally lose their nucleus and are then passed out for circulation in the blood.

The *average life of a red blood cell* is about 120 days. The cells then wear out. They are disintegrated in the reticulo-endothelial system, principally in the spleen and liver. The *globin* of the haemoglobin is broken down into amino-acids to be used as *protein* in the tissues and the iron in the haem is removed for use in the formation of future red blood cells. The rest of the haem is converted into *bilirubin*, the yellow pigment, and *biliverdin*, the green one seen in the colour changes of reduced haemoglobin in a *bruise.*

When bleeding occurs red cells, with their oxygen-carrying haemoglobin, are lost. In moderate haemorrhage these cells will be replaced during the following weeks, provided a balanced diet, containing adequate iron, is taken; but if the percentage of haemoglobin falls to 40 or below, a blood transfusion may be needed. (*See also* Clinical Notes, page 176.)

Haemoglobin is a complex protein rich in iron. It has an affinity for oxygen, and combines with it forming *oxy-haemo-globin* in the red cells. By means of this *function* oxygen is carried to the tissues from the lungs.

The amount of haemoglobin present in normal blood is about 15 g per 100 ml blood and this amount is usually called '100 per cent'. Anything over 90 per cent is considered normal.

In many forms of anaemia the amount of haemoglobin present in the blood is diminished. In some severe forms it may fall below 30 per cent, that is, 5 g per 100 ml. As haemo-globin contains the iron necessary to combine with oxygen, it will readily be understood that these patients present symp-toms of deficient oxygen such as breathlessness, often one of the first indications of iron deficiency anaemia.

Blood Groups. Substances in the plasma, called agglutinins, will, if blood of an incompatible group is transfused, cause clumping and haemolysis (breakdown) of the red cells.

Blood grouping and tests of incompatibility are carried out in order to ensure a high degree of safety before a *blood trans-fusion* is given. The Landsteiner ABO system is based on the agglutinin content of the blood. The four main groups designated are:

Group AB representing $3\cdot0\%$ of people in Great Britain
,,   A   ,, $42\cdot0\%$,,   ,,   ,,   ,,   ,,
,,   B   ,, $8\cdot5\%$,,   ,,   ,,   ,,   ,,
,,   O   ,, $46\cdot5\%$,,   ,,   ,,   ,,   ,,

In addition there are a number of Landsteiner sub-groups and the Rh or Rhesus factor in blood is of importance in the newly-born when the blood of the fetus may be incompatible with that of the mother.

In considering *donors of blood*:

Group AB may give blood to AB
   ,,   A to A and AB
   ,,   B ,, B ,, AB
   ,,   O is a universal donor for all groups.

Recipients:

Group AB is a universal recipient
   ,,   A may receive blood from groups A and O
   ,,   B ,, ,, ,, ,, ,, B ,, O
   ,,   O from O.

It is customary to transfuse blood of the same group as the patient, and only in emergency to give the blood of a universal donor.

White Blood Cells, which are transparent and not coloured, are larger and fewer than the red. There are from 6,000 to 10,000 (with an average of 8,000) in each cubic millimetre of blood. They are classified as follows:

Granulocytes or *polymorphonuclear cells* form almost 75 per cent of the total white cell count; they are formed in the red marrow of bone. These cells contain a many-lobed nucleus and the protoplasm of the cells is granular, hence the term granular cell or granulocyte.

Deficiency of granulocytes is described as *granulocytopenia*. Complete absence of them, *agranulocytosis*, may arise when certain drugs are given, including some of the antibiotics. Therefore under these conditions, frequent blood examinations are made in order to detect this condition as soon as possible.

Staining. When a drop of blood is placed on a slide and two stains are added in order to make a blood count, cells in this group are typed according to their manner of staining.

Neutrophil cells form the majority; these stain with neutral dyes, or a mixture of the acid and alkaline stains, and appear *purple*.

Eosinophil cells. A very few cells form this group; they take the acid (eosin) stain and appear *red*.

Basophil cells take the basic dyes and stain blue.

Lymphocytes form about 25 per cent of the total white cell count (*see* page 169). These cells are developed in the lymph glands, the spleen, liver, and lymphatic tissue as well as in the bone marrow. They are non-granular cells and have no power of amoeboid movement. They are subdivided into small and large lymphocytes. In addition a few larger cells (about 5 per

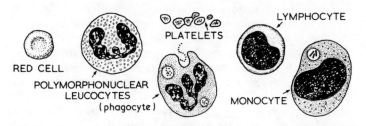

Fig. 10/2.—White Blood Cells. A Red Cell and Platelets show Relative Size

cent) are described as *monocytes*; these cells are capable of amoeboid movement and are phagocytic in action.

Function of the White Cells. The *granulocytes* and *monocytes* play a very important part in protecting the body from micro-organisms. By *phagocytic action* (*phago*—I eat) they ingest living bacteria. As many as 10 to 20 micro-organisms may be seen in a granulocyte under microscopic examination. When performing this function they are called *phagocytes*. By their power of amoeboid movement they can move freely in and out of the blood vessels and wander about in all parts of the body. In this way they can:

Surround any area which is infected or injured,
Take in living organisms and destroy them (ingestion),
Remove other materials, such as bits of dirt, splinters of wood, catgut sutures, etc., by a similar process, and

In addition, the granulocytes possess a protein-splitting ferment which enables them to act on living tissue, break it
down, and remove it. In this way diseased or injured tissue
can be removed and healing promoted.

*As the result of the phagocytic action of the white blood cells
inflammation may be entirely arrested.* When the activity does
not proceed to complete resolution, pus may be formed. *Pus*
consists of the dead bodies of friends and foes—phagocytes
killed in the battle against the invading germs are called
pus cells. Many dead germs also are present in pus and in
addition there is a considerable amount of liquefied tissue.
As the fight proceeds, if the white cells overcome the invading organisms, eventually all signs of destruction will be
removed, living and dead bacteria, pus cells and liquefied
tissue all being removed by the healthy granulocytes acting as
phagocytes.

Of the *function of the lymphocytes* less is known. They have
no power of amoeboid movement, they float in the blood
stream and are also found in lymphatic tissue in all parts of
the body. They do not ingest bacteria, but it is thought that
they make valuable antibodies to protect the body against
chronic infection and to maintain some degree of immunity to
all infections.

Leucocytosis is the term used to describe increase in the total
number of white cells in the blood when the increase exceeds
10,000 per cubic millimetre.

Leucopenia means decrease in the white count to 5,000 or less.

Lymphocytosis—increase in the number of lymphocytes.

Agranulocytosis—a marked decrease in the number of
granulocytes or polymorphonuclear cells.

Blood Platelets or Thrombocytes are small cells about one-
third the size of a red blood cell. There are 300,000 of them in
each cubic millimetre of blood. They play an important part
in the control of bleeding after injury and in the clotting of
blood.

Summary of the Number of Blood Cells in each cubic millimetre of Blood

The Normal blood count or the number of cells per cubic millimetre of blood is approximately:

Red Cells   4,500,000 to 5,500,000   Average   5,000,000

White Cells   6,000 to 10,000   Average   8,000
Made up as follows:

	per cent	Average per cent
Granulocytes:		
Neutrophil cells	60 to 70	66
Eosinophil cells	1 to 4	3
Basophil cells	$\frac{1}{2}$ to 2	1
Lymphocytes (large and small) . .	20 to 30	25
Monocytes . . .	4 to 8	5
	Total	100

Platelets   250,000 to 500,000   Average   350,000

Blood Plasma is a straw-coloured fluid, slightly alkaline in reaction. The composition of plasma and the list of substances contained in it is given on page 163.

  Functions of plasma. Plasma acts as the medium for the transmission of nutriment, salts, fats, glucose, and amino-acids to the tissues; and as the medium for carrying away waste materials—urea, uric acid, and some of the carbon dioxide.

  Plasma proteins. *Albumin.* There are normally 3 to 5 g of albumin in each 100 ml of blood. It has three functions:

(1) it is responsible for the osmotic pressure which maintains the blood volume,

(2) many special substances are carried in combination with the albumin, and

(3) it provides protein to the tissues.

Globulin. There are normally 2 to 3 g of globulin in each 100 ml of blood. Globulin is much more variable than albumin in composition and really comprises a very large number of different proteins. It is less important in providing osmotic pressure than albumin but more important in other ways; for instance all the protective antibodies are globulins.

Fibrinogen is essential for blood coagulation (*see* below).

The reaction of blood plasma. Blood is always alkaline; the degree of alkalinity depends on the hydrogen-ion concentration and this is expressed as the *p*H of blood.

The *p*H of 7 —— represents a neutral solution.

The *p*H from 7 to 1 —— an acid solution.

The *p*H from 7 to 14 —— an alkaline solution.

It will be seen that the figure *p*H 7 is a neutral solution. Blood is always slightly alkaline—the *p*H of blood is 7·35–7·45. This figure is constantly maintained; only very slight variation on either side is compatible with life. The maintenance of the constant degree of alkalinity of the blood therefore is most important and this is controlled by the following factors:

The *elimination of carbon dioxide* (which is an acid gas) from the lungs.

The *excretion of acids* in the urine.

The *alkaline reserve* property of the blood, which depends on the presence of sodium bicarbonate in the plasma, acting as what is described as a *buffer substance* and preventing reduction of the alkalinity of the blood by acids resulting from metabolism.

The Coagulation of Blood. When blood has been shed, it quickly becomes sticky and soon sets as a red jelly. This jelly or clot contracts or shrinks, and a straw-coloured fluid called *serum* is squeezed out from it.

If shed blood is microscopically examined, very fine threads will be seen, the insoluble *fibrin* threads formed from the *fibrinogen* in the blood plasma *by the action of a ferment thrombin*. These threads entangle the blood cells and together with them form the clot. If shed blood is collected in a test tube, the clot will eventually float in the serum.

The clotting of blood is a complicated process, and several factors are necessary to bring it about. As already stated the ferment *thrombin* is instrumental in converting *fibrinogen* into *fibrin threads*. *Thrombin* is not present in normal unshed blood but its precursor *prothrombin* is present and is converted into the active ferment thrombin by the action of thrombokinase. *Thrombokinase* or *thromboplastin* is an activating agent liberated on injury to the blood cells, it is thought largely by injury to the blood platelets, which, provided that *calcium salts* are present in the blood, will convert prothrombin into thrombin so that clotting can take place.

To produce a clot therefore four factors are necessary:

Calcium salts, normally present in blood,

Cell injury which liberates thrombokinase,

Thrombin formed from prothrombin in the presence of thrombokinase, and

Fibrin formed from fibrinogen in the presence of thrombin.

The *process of clotting* may be expressed by the formula:

Prothrombin + calcium + thrombokinase	= *Thrombin*
Thrombin + fibrinogen	= *Fibrin*
Fibrin + blood cells	= *Clot*

Prothrombin is made in the liver. Vitamin K is necessary for its production.

Coagulation is hastened (*a*) by heat a little higher than the body temperature, (*b*) contact with rough material, such as the roughened edge of a damaged blood vessel or a surgical dressing.

It is *retarded* (*a*) by cold, (*b*) by being kept in a vessel coated with paraffin wax because blood needs to be in contact with a surface that can be wet by water before it will clot, and paraffin is not a water-wetting surface, (*c*) by the addition of potassium citrate or sodium citrate, which removes the calcium salts normally present.

Clinically a *thrombus* is a clot formed in the circulation; the condition is one of thrombosis. A femoral thrombosis may occur after an operation; a clot in a coronary artery causes *coronary thrombosis*. When a portion of a clot becomes detached and enters the circulating blood it is called an *embolus*. If this were to pass through the heart and enter the lungs by one of the pulmonary arteries, a small or large vessel might be blocked by it, constituting a *pulmonary embolism*.

Summary of the Functions of Blood

(1) To act as the transport system of the body, conveying all chemical substances, oxygen and nutrients required for the nourishment of the body in order that its normal functions may be fulfilled, and carrying away carbon dioxide and other waste products.

(2) The red cells convey oxygen to the tissues and remove some of the carbon dioxide.

(3) The white cells provide many of the protective substances and by phagocytic action some of the cells protect the body against bacteria.

(4) The plasma distributes proteins needed for tissue formation; it services the tissue fluid by which all cells receive nourishment; and forms the vehicle by which waste matter is conveyed to the various excretory organs for elimination.

(5) The internal secretions, hormones, and enzymes are conveyed from organ to organ by means of the blood.

All tissues need an adequate supply of blood, which depends on a normal arterial blood pressure being maintained (*see* below). In the lying position the blood pressure in the body is level, but when sitting or standing the blood to the brain has to be pumped uphill.

The brain in particular needs a continuous adequate supply of blood; if deprived of blood for longer than 3 to 4 minutes, irreversible changes take place and some brain cells die. Therefore in cardiac arrest from any cause (*see* page 161), urgent immediate treatment is essential to get the heart acting again. Even in a simple faint, such as may be caused by emotional or physical strain a fall in blood pressure diminishes the blood supply to the brain. Therefore it is essential to lower the head of the subject, either by pressing it forward between his knees if sitting, or preferably to have him lying on the floor.

BLOOD PRESSURE

Arterial Blood Pressure is the force of pressure which the blood is exerting against the walls of the blood vessels in which it is contained. This pressure varies during the cardiac cycle (*see* page 154).

During ventricular systole, when the left ventricle is forcing blood into the aorta the pressure rises to a peak, *systolic pressure*. During diastole the pressure falls, the lowest value it reaches being called *diastolic pressure*.

Systolic blood pressure is produced by the heart muscle which drives the contents of the ventricle into the already stretched arteries. During diastole the arteries are kept partly distended because the *peripheral resistance of the arterioles* prevents all the blood running off into the tissues. Thus the blood pressure depends partly on the force and volume of the blood pumped by the heart and partly on the contraction of the muscles in the walls of the arterioles. This contraction is maintained by vasoconstrictor nerves which are controlled by the *vasomotor centre* in the medulla oblongata of the brain.

The vasomotor centre adjusts the peripheral resistance to maintain the blood pressure relatively constant. It changes slightly in physiological variations of exertion as in exercise, with mental changes of anxiety and emotion, in sleep and when eating. For this reason the blood pressure is always taken when a person is relaxed, resting and preferably recumbent.

In Measuring Arterial Blood Pressure an instrument called a sphygmomanometer is used. The upper arm is encircled by an inflatable rubber bag contained in a cuff which is connected to a pressure pump and manometer. By pumping, the pressure in the bag is rapidly raised to 200 mm Hg which is sufficient to obliterate completely the brachial artery so that no blood comes through, and the radial pulse disappears. The pressure is then lowered to a point where the pulse can be felt or, more exactly, when, by using a stethoscope, the pulsation of the brachial artery at the bend of the elbow can be distinctly heard. At this point the pressure shown on the column of mercury in the manometer is considered to be the *systolic pressure*. The pressure on the brachial artery is then gradually reduced until the heart sounds or arterial pulse beats can be distinctly heard or felt, and the point at which the sounds begin to fade is generally accepted as the *diastolic pressure*.

The difference in pressure between systole and diastole is called the **pulse pressure** and is normally from 30 to 50 mm Hg. The lower limit of systolic pressure in the normal adult is estimated at approximately 105 mm Hg, and the upper limit at 150. In women the blood pressure is from 5 to 10 mm Hg lower than in men.

Normal Blood Pressure Range (in mm Hg)

	Diastolic	Systolic
In infancy the blood pressure is	50	70 to 90
In childhood	60	80 to 100
During the adolescent period	60	90 to 110
In the young adult	60 to 70	110 to 125
As age advances it is increased	80 to 90	130 to 150

Factors Maintaining Blood Pressure

The pumping force of the heart, already mentioned on p. 154.

The quantity of circulating blood. It is necessary to fill any system of tubes to capacity in order to develop pressure. As the blood vessel walls are elastic and distensible these must be overfilled before any degree of pressure can be effected. Loss of blood, as in haemorrhage, will result in a fall of pressure. The administration of fluid such as blood plasma or saline will cause the pressure to rise again.

The viscosity of blood. Blood derives its viscosity from plasma proteins and from the number of corpuscles contained in the blood stream. Any change in these two factors will alter the blood pressure. For example in anaemia the blood corpuscles are decreased in number and consequently the pressure would be lower if the heart and vasomotor system did not overact to compensate.

The amount of friction exerted by fluid on the walls of tubes through which it is flowing varies according to the viscosity of the fluid. The more concentrated the fluid the greater will be the force required to drive it through the vessels.

The elasticity of the walls of the blood vessels. The pressure is greater in arteries than in veins because the muscular coat of the arteries is more elastic than that of the veins.

The peripheral resistance. This is the resistance offered by the friction of the blood flowing in the vessels. The main resistance to the flow of blood in the systemic circulation lies in the arterioles and the greatest fall in pressure will occur here. The arterioles also 'smooth the pulsation out' of the blood pressure and blood flow so that it is not seen in the capillaries and veins.

The Velocity of the Blood. The rate at which the blood is flowing depends on the size of the bed provided by the vessels or groups of vessels. The *blood in the aorta* is moving rapidly. It slows down in the arteries and becomes very slow in the capillaries.

Some pressure can be noted as the return blood reaches the larger vessels (veins) nearing the heart.

In *the capillary bed,* or capillary 'lake' as it is sometimes called, the blood is flowing through a very great number of extremely minute vessels; the actual cross-section of the area provided by these vessels is about 600 times that of the aorta.

This widening of the area through which the same amount of blood flows results in a marked slowing of the stream. It is here, in this very slow stream, that the interchange of gases, absorbed food substances, and waste products takes place between the red blood corpuscles and plasma in the capillaries and the fluids and cells in the body tissues.

As the blood is collected by veins the rate of flow is increased again and the blood flowing through the lumen of the inferior and superior venae cavae together is as rapid as the stream in the aorta. In order to maintain the circulation, the blood reaching the heart must be of the same volume as the blood leaving the heart. *Blood pressure in the veins* is low and other factors assisting the flow of blood back to the heart include:

The movements of the skeletal muscles exerting pressure on the veins.

The movements produced by breathing, in particular by the rise and fall of the diaphragm which acts as a pump.

A *suction action* exerted by the atria, empty during diastole, attracts blood from the veins to fill them.

The *arterial blood pressure* which, although greatly decreased by the arterioles and capillaries, is still sufficient to drive the blood onward.

Clinical Notes

Anaemia is a deficiency in the number of the red blood cells or in the amount of haemoglobin contained in them. There are a number of varieties of anaemia and a complex classification of this disorder of the blood exists, but for the purposes of this brief note two types only are mentioned.

Iron deficiency anaemia may be due to severe haemorrhage arising as the result of injury or disease; to constant slight losses of blood as in peptic ulcer and in hiatus hernia; to nutritional deficiency which may be the result of eccentric habits of eating, ignorance in budgeting for the family or in cooking, or lack of food due to poverty. Further, unless essential chemical factors are present in the diet, such as Vitamin B_{12} (cyanocobalamin) present in red meats and liver, the process of production of the red blood cells is incomplete and they are not mature.

Pernicious anaemia (Addison's anaemia) occurs when there is failure of the absorption of Vitamin B_{12} mentioned above. This is due to a defect of the stomach which fails to secrete a special substance, normally secreted by the wall of the stomach, known as *the intrinsic factor of Castle*, essential for the absorption of Vitamin B_{12}.

The *basic symptoms of anaemia* are those due to lack of the oxygen-carrying factor in blood and are characterized by breathlessness, listlessness, and a feeling of tiredness, and loss of appetite, with pallor of the skin. Iron-deficiency anaemia responds to the regular administration of iron; pernicious anaemia to regular intramuscular injections of Vitamin B_{12} (Cytamen is one of proprietary preparations given).

Leucocytosis, leucopenia and *agranulocytosis* are mentioned in the text, page 168.

Haemophilia is a familial disorder of the blood with deficiency in the clotting mechanism, so that a patient may bleed severely after a very slight injury and a dangerous degree of bleeding may follow tooth extraction. The disorder is transmitted through the females, who are carriers, to the males, who are the haemophiliacs.

The *phagocytic action* of white blood cells is mentioned on pages 167–8, and the gravity of the reduction of haemoglobin in the red cells on page 165.

Hypertension implies a blood pressure raised above normal, but it is difficult to define an average normal blood pressure for any group, though at the age of sixty the figure 160/90 is often given. Nevertheless many, apparently normal, people have a pressure higher than this figure. A range of 40 to 50 generally exists between the systolic and diastolic pressure. A diastolic pressure of over 130 mm Hg is a serious degree of hypertension.

The cardiovascular system is carefully examined in any subject with hypertension. There may be some cardiac involvement, changes in the arteries may arise; for example, a cerebral artery may give way, with resulting 'stroke', or a retinal haemorrhage may occur, or changes arising in the kidneys may disturb the renal function.

Hypotension, a low blood pressure, can be physiological in health, during rest, after fatigue and in some elderly people. It occurs as a symptom in myxoedema, and is met in some forms of non-malignant thyroiditis as in Hashimoto's disease, when it will usually respond to the administration of thyroid extract.

Chapter 11

THE PRINCIPAL BLOOD VESSELS

There are several kinds of blood vessels. *Arteries and Arterioles*, which convey blood away from the heart, always carry oxygenated blood, the exception being the pulmonary arteries which carry venous blood.

Venules and Veins carry blood towards the heart and, except the pulmonary veins, always carry deoxygenated blood.

Capillaries are very minute blood vessels in which arterioles terminate and venules begin. They form a delicate network of vessels which ramify in most parts of the tissues of the body.

Certain arteries, such as those carrying blood to the brain, and some of the vessels of the lungs, liver, and spleen, do not terminate in ordinary capillaries.

The Structure of Blood Vessels. *Arteries* are composed of three coats:

Outer fibrous and connective tissue coat, *tunica adventitia*,

Middle muscular and elastic coat, *tunica media*, and

Inner endothelial coat, *tunica intima*.

The outer coat is protective. The middle layer is strong; it holds the vessel open and by the state of contraction of the muscle fibres exerts steady pressure on the blood.

The inner endothelial coat is very smooth, being lined by a single layer of flat pavement cells.

The middle coat of the aorta and the larger arteries contains

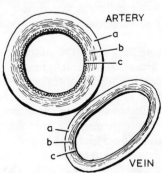

FIG. 11/1.—STRUCTURE OF ARTERY AND VEIN

A. Adventitia.
B. Media.
C. Intima.
Note the thick muscular coat of the artery.

177

a large quantity of elastic fibre and less muscular tissue, as these arteries require to be very distensible. The smaller arteries and arterioles contain relatively more muscle tissue, as their walls must be readily adapted, by the vasomotor control, to the needs of the body.

The thick walls of larger arteries are themselves supplied with blood by a special system of tiny vessels, known as the *vasa-vasorum*; they have also a nerve supply of slender nerve filaments embracing the walls of the vessels.

Veins are composed of the same three layers as the arteries, but the middle muscular layer is thinner, less firm, more collapsible, and much less elastic than the arteries. The veins in the limbs where the blood travels against gravity have valves arranged so as to allow the blood to flow towards the heart, but not in the opposite direction. These valves are crescent-shaped folds composed of the inner lining of the vein, endothelium, strengthened by a little fibrous tissue. The folds are opposite one another; their free edge is in the direction in which the blood is flowing. When distended with blood the valves give a knotted appearance to the vein.

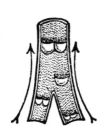

FIG. 11/2.—A VEIN OPENED TO SHOW THE FOLDS WHICH FORM THE VALVES

Capillaries are the minute vessels in which the arteries terminate. As the arterioles get smaller and smaller, the three coats gradually disappear until, when the fine hair-like capillary vessels are formed, these consist of one layer, the inner endothelial coat of the arteries. The extreme thinness of these vessels permits the transudation of lymph, which forms the tissue fluid and brings water, valuable salts, and nutriment to the cells, and by the interchange of gases between the capillary vessels and tissue cells, supplies oxygen, and carries away waste matter including carbon dioxide.

The capillaries, therefore, perform a very important function, as they distribute the substances to the tissues which enable the various processes of the body to go on.

The Composition of the Blood Varies in the arteries and veins. *Arterial blood* contains oxygen and is *bright scarlet* in colour because the haemoglobin is combined with oxygen. If an artery is cut across, this bright red blood will be seen to spurt out in jets corresponding with the heart beat.

Venous blood is darker and *purplish* in colour, as much of the oxygen has been given up to the tissues. If a vein is cut across, the blood flows out in an even stream. *Blood in the capillaries* is continually changing in composition and colour, due to the interchange of gases taking place. Capillary bleeding is recognized by the blood oozing smartly on to the surface.

NAMES AND POSITIONS OF THE CHIEF ARTERIES

The Aorta is the main artery in the body. The part situated in the thorax is known as the *thoracic aorta*. The aorta leaves the left ventricle of the heart where its opening is guarded by the aortic valve. It then arches over the base of the heart, *as the arch of the aorta*, reaching as high as the manubrium sterni. Three branches arise from this arch. One on the right, the *innominate* artery, is about 5 cm (2 inches) long and divides into *right common carotid* and *right subclavian* arteries. Two branches arise from the left side of the arch, the *left common carotid* and the *left subclavian* arteries.

From *the arch of the aorta* the vessel passes through the thorax as the *thoracic aorta*, passes behind the diaphragm, and becomes the *abdominal aorta*. The aorta gives off branches to supply the thoracic and abdominal viscera (*see* Fig. 11/3).

The Abdominal Aorta gives off a number of important branches.

The *coeliac axis* arises just beneath the diaphragm, and divides into three branches, the *hepatic*, *gastric*, and *splenic* arteries, to supply the liver, stomach, pancreas, and spleen.

The *mesenteric* arteries, superior and inferior, supply the mesentery and intestines.

The *renal* arteries supply the kidneys.

The *testicular* arteries in the male and the *ovarian* in the female supply the testes and the ovaries respectively.

In front of the fourth lumbar vertebra *the abdominal aorta* divides into the right and left *common iliac* arteries. These then divide into the right and left *internal* and *external iliac* arteries. The internal iliac enters the pelvis to supply the organs there.

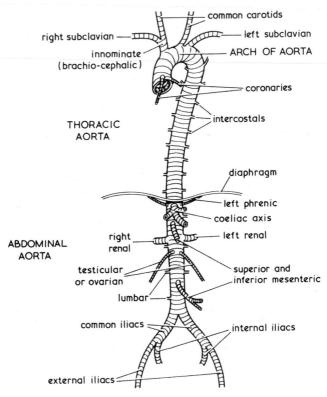

FIG. 11/3.—THE AORTA AND ITS MAIN BRANCHES

The external iliac passes under the inguinal ligament, enters the thigh, and becomes the *femoral* artery.

The Common Carotid Artery. The common carotid artery ascends in the neck and divides into the *internal* and *external carotid* arteries at the level of the thyroid cartilage.

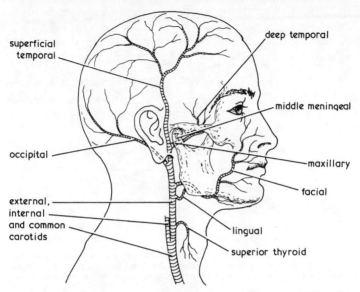

FIG. 11/4.—BRANCHES OF THE EXTERNAL CAROTID SUPPLYING THE FACE AND HEAD. THE INTERNAL CAROTID PASSES INTO THE SKULL (*see* FIG. 11/5, PAGE 182)

The *internal carotid* artery ascends without giving off any branches in the neck, passes through the carotid canal in the temporal bone, enters the skull, and divides into *ophthalmic* and *anterior* and *middle cerebral* arteries.

The *external carotid* artery divides into three main branches to supply the outer side of the cranium and face.

The *facial* artery passes over the mandible near its angle, divides at the angle of the mouth into labial branches and ascends to supply the nose and cheek, finally ending at the medial corner of the eye.

The *temporal* artery which ascends at the side of the head can be felt pulsating where it lies superficial to the temporal bone in front of the ear. It gives branches to the side of the head.

The *occipital* artery passes to the back of the head and divides into branches to supply this part.

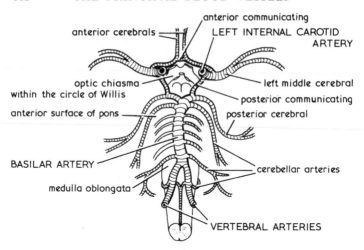

FIG. 11/5.—THE CIRCLE OF WILLIS. THE TWO VERTEBRAL ARTERIES AT THE BACK AND THE TWO INTERNAL CAROTID ARTERIES IN FRONT UNITE TO FORM THE CIRCLE OF WILLIS, FROM WHICH ARTERIES ARISE TO SUPPLY THE BRAIN

The *maxillary* artery passes deep to the neck of the mandible into the cheek and supplies the muscles of mastication. It also gives off the important *middle meningeal artery* which passes up through the base of the skull into the cranial cavity. This artery is sometimes the site of an 'extradural' haemorrhage following a fractured skull as it runs in a groove on the inner aspect of the skull.

Circle of Willis. The *vertebral* arteries arise from the sub-clavian arteries, and, passing up the neck through the foramina in the transverse processes of the cervical vertebrae, enter the skull, through the foramen magnum, uniting to form the *basilar* artery. This gives off the right and left *posterior cerebral* arteries. Branches from these pass forwards and anastomose with the middle and anterior cerebral arteries to form the *circle of Willis*.

The Subclavian Artery and its Terminations. The subclavian artery arising from the aorta passes over the first rib, which it grooves; then beneath the clavicle to enter the axilla, where

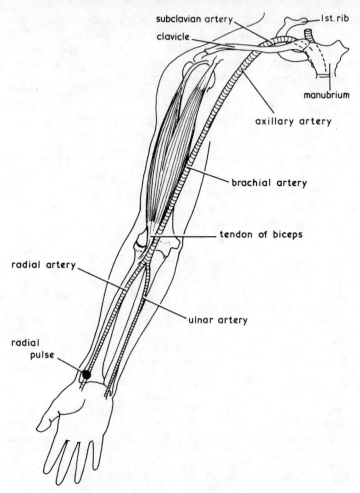

FIG. 11/6.—THE MAIN ARTERIES OF THE RIGHT UPPER LIMB

it becomes the *axillary* artery. At the lower boundary of the axilla it becomes the *brachial* artery, which runs down the arm at the medial side of the biceps muscle, to divide at the bend of the elbow into *radial* and *ulnar* arteries.

In the *cubital fossa* (front of the bend of the elbows), the tendon of biceps may be clearly felt in the middle line, if the elbow joint is flexed. Just medial to this, the large *brachial artery* is easily palpated. This is the usual site for recording the blood pressure and affords a most useful alternative to the radial artery for recording the pulse rate as the vessel at this point is larger.

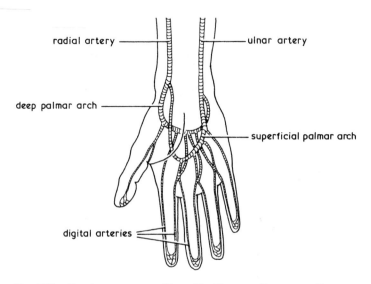

FIG. 11/7.—THE ARTERIES OF THE HAND. THE DEEP AND SUPERFICIAL PALMAR ARCHES

The *radial artery* passes down the radial side and the *ulnar artery* down the ulnar side of the forearm, supplying blood to the structures of this region. Passing over the front of the wrist, terminations of these arteries form the *deep* and *superficial palmar arches* in the hand. These give off *palmar* and *digital branches* to the hand and fingers.

The Femoral Artery and its Terminations. The femoral artery passes down the medial aspect of the thigh and in the lower third passes behind the femur and behind the knee joint, where

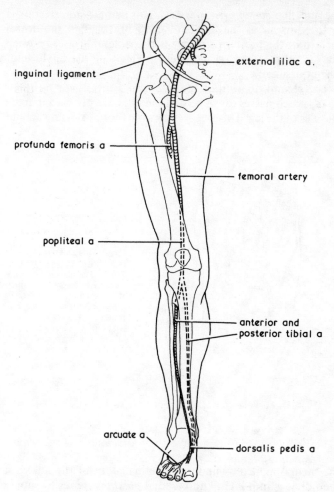

FIG. 11/8.—THE MAIN ARTERIES OF THE RIGHT LOWER LIMB

it becomes the *popliteal* artery. It then divides into two main arteries to supply the leg.

The *anterior tibial* artery lies in the anterior compartment of the muscles of the leg, and passing over the bend of the ankle becomes the *dorsalis pedis* artery. This supplies the structures in the dorsum of the foot and gives branches to the dorsal surface of the toes. This artery is palpable, mid-way between the lateral and medial malleoli, in front of the ankle joint when dorsiflexed.

The second division of the popliteal artery is the *posterior tibial*, which passes down behind the tibia, deeply placed in the muscles of the leg. This artery enters the foot by passing behind

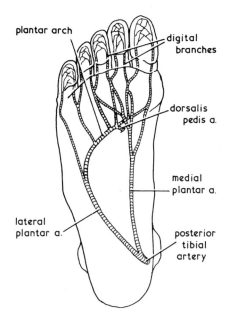

FIG. 11/9.—ARTERIES OF THE SOLE OF THE FOOT. THE POSTERIOR TIBIAL ARTERY PASSES BEHIND THE INTERNAL MALLEOLUS AND DIVIDES INTO EXTERNAL (LATERAL) AND INTERNAL (MEDIAL) PLANTAR ARTERIES

the medial malleolus under the retinaculum of the ankle. It then divides into lateral and medial *plantar* arteries to supply the structures in the sole of the foot.

Arterial grafting is undertaken for the relief of arterial obstruction. Nurses should be familiar with the course of the blood supply beyond or below the lesion in order to make accurate observations of the temperature and colour of the limb, be aware of and report any sudden cooling or mottling of the skin, be familiar with the position of the peripheral pulses, capable of palpating them, and reporting any sudden changes.

Knowledge of the reason for the period of bed-rest ordered and the measures to be taken should any complication, oedema for example, arise, ought to be part of her equipment.

THE PRINCIPAL VEINS

Veins carry blood towards the heart. They begin as small vessels formed by the union of capillaries. These small veins unite and become larger veins, eventually forming the venous trunks which increase in size as they near the heart. Veins are more numerous and larger than arteries.

The Deep Veins or Venae Comites accompany the main arteries and are named after them; some arteries have two accompanying veins.

The Superficial Veins lie immediately beneath the skin, and communicate with the deep veins at certain points, before the great venous trunks reach the heart.

Veins of the Thorax. The brachiocephalic veins formed by the union of the subclavian and internal jugular veins unite behind the first costal cartilage to form the superior vena cava. The right brachiocephalic is shorter than the left vein. The brachiocephalic veins receive the blood from the head and upper limbs and in addition receive veins from the upper part of the thorax including the mammary veins.

The *Azygos group of veins* receives veins from the thorax including the bronchial veins, and the azygos vein enters the superior vena cava.

The *Superior Vena Cava* formed by the union of the two brachiocephalic veins is about 7·5 cm (3 inches) long. It receives the blood from the head, neck, both upper limbs and the anterior walls of the thorax and empties its contents into the upper part of the right atrium of the heart.

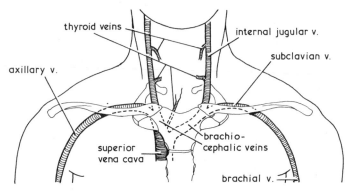

FIG. 11/10.—FORMATION OF THE SUPERIOR VENA CAVA BY THE UNION OF THE TWO BRACHIOCEPHALIC VEINS

Deep Veins of the Upper Extremity. In *the upper limb* there are the *radial* and *ulnar* veins in the forearm; these unite at the elbow and become the *brachial* vein; this becomes the *axillary*, and finally the *subclavian* vein. The subclavian vein from each side then unites with the *internal jugular* vein from the head forming the right and left *brachiocephalic* veins, and these two brachiocephalic veins unite to form the *superior vena cava* (*see* Fig. 11/10).

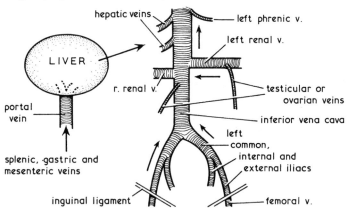

FIG. 11/11.—THE INFERIOR VENA CAVA AND MAIN TRIBUTARIES (Cf. FIG. 11/3, PAGE 180)

Veins of the Pelvis and Abdomen. The *femoral vein* passes from the lower limb beneath the inguinal ligament to enter the pelvis where it becomes the *external iliac vein*. Near the sacro-iliac joint it unites with the *internal iliac vein* which drains the blood from the organs in the pelvis. This union of external with internal iliac becomes the *common iliac vein*, right and left common iliac veins unite towards the right side of the fifth lumbar vertebra to become the inferior vena cava.

The *Inferior Vena Cava* receives many tributaries as it passes up through the abdomen to convey the blood from the parts below the diaphragm to the heart. It receives the lumbar veins which drain the posterior abdominal wall, the testicular or ovarian veins, the renal and suprarenal, the inferior phrenic and the hepatic veins. The respiratory movements of the diaphragm form part of the 'muscle pump' attracting venous blood from the lower limbs towards the heart.

Deep Veins of the Lower Extremity. In *the lower limb* the *anterior* and *posterior tibial* veins unite to become the *popliteal*, which then becomes the *femoral*, and finally becomes the *external iliac*. The internal and external iliac veins unite to form the *common iliac* vein. The right and left common iliac veins unite and the *inferior vena cava* is formed (*see above and* Fig. 11/11).

The Veins of the Head and Neck. The blood from the brain drains in the interior of the skull into channels formed by the dura mater, called **venous sinuses** (*see* Fig.11/12).

The *superior longitudinal* or *sagittal sinus* which corresponds in position to the upper border of the falx cerebri receives blood from the brain and the *inferior sagittal* drains the falx cerebri and from the nearby brain tissue.

The *straight sinus* is situated between the falx cerebri and the tentorium cerebelli.

Two *transverse sinuses* lie close to the skull, receiving blood from other sinuses, and, passing to an opening in the skull known as the jugular foramen, become, in the neck, the right and left internal jugular veins.

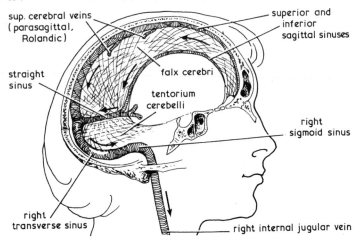

FIG. 11/12.—THE PRINCIPAL VENOUS SINUSES OF THE DURA MATER

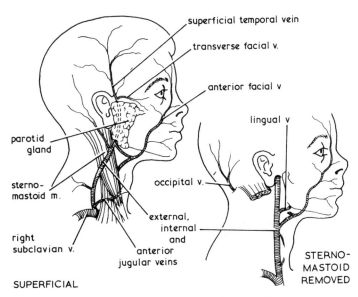

FIG. 11/13.—THE VEINS OF THE HEAD AND NECK

Other sinuses lie deep in the skull including the *cavernous sinuses* which lie one on each side of the sphenoid bone. These drain the regions of the orbit, nose and cheek as well as parts of the cerebral hemispheres. Infection of the face in these regions may cause thrombosis of the cavernous sinus which is a serious condition.

The Internal Jugular Vein lies *deep* in the neck; it contains the blood which has drained from the interior of the skull and it receives also the lingual, facial and thyroid veins. The internal jugular vein unites with the subclavian vein on each side to form the right and left brachiocephalic veins. These brachiocephalic veins unite to form the superior vena cava (*see* Fig. 11/10).

The External Jugular Vein is a *superficial vein* formed slightly behind and below the ear by the union of veins draining the regions of the side of the face and the ear. It enters the subclavian vein. Another superficial vein lying at the front of the neck—the *anterior jugular vein*—drains this area and joins the external jugular (*see* Fig. 11/13).

The Superficial Veins of the Upper Extremity begin as a network of small veins in the hands. Those from the palm drain into the median vein, those from the medial aspect of the dorsum into the basilic and those from the lateral aspect into the cephalic vein.

The *median vein* runs up the anterior aspect of the forearm and below the elbow divides into two veins which enter the basilic and cephalic veins. These latter two veins are connected by the *median cubital vein* (*see* Fig. 11/14) which crosses the cubital fossa and is commonly used for venepuncture.

The *basilic vein* runs up the medial aspect of the forearm and pierces the deep fascia in the upper arm. It is continued on as the (deep) brachial vein which becomes the axillary vein.

The *cephalic vein* runs up the lateral aspect of the forearm and arm until it pierces the deep fascia near the shoulders to pour its contents into the axillary vein.

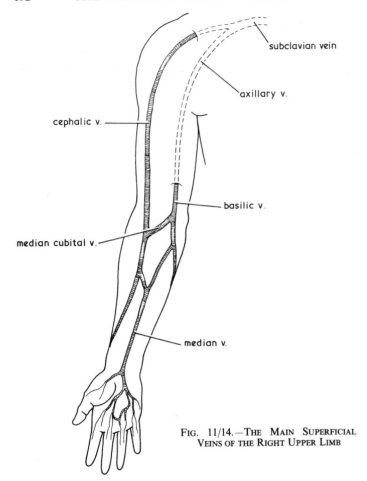

subclavian vein

axillary v.

cephalic v.

basilic v.

median cubital v.

median v.

FIG. 11/14.—THE MAIN SUPERFICIAL VEINS OF THE RIGHT UPPER LIMB

Veins of the Thorax. The brachiocephalic veins formed by the union of the subclavian and internal jugular veins unite behind the first costal cartilage to form the superior vena cava. The right brachiocephalic is shorter than the left vein. The brachiocephalic veins receive the blood from the head and upper limbs and in addition receive veins from the upper part of the thorax including the mammary veins.

The Superficial Veins of the Lower Extremity. The long *saphenous* vein is the largest. It commences on the medial aspect of the dorsum of the foot, receiving tributary veins from this region; then passes up along the medial aspect of the leg, behind the knee, to come forward again and finally pierce the deep fascia at the saphenous opening (*see* Fig. 11/15) to enter the femoral vein within the femoral sheath. It receives tributary veins in its whole course, and is accompanied by numerous lymphatic vessels.

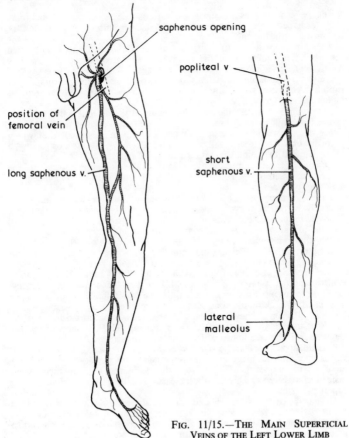

saphenous opening

popliteal v

position of
femoral vein

long saphenous v.

short
saphenous v.

lateral
malleolus

FIG. 11/15.—THE MAIN SUPERFICIAL
VEINS OF THE LEFT LOWER LIMB

The short or *small saphenous* vein commences on the lateral border of the foot. It passes behind the lateral malleolus and along the middle of the calf of the leg to the knee. It receives tributaries from the foot and back of the leg, and finally pierces the fascia in the popliteal region to join the (deep) popliteal vein.

There are communications between superficial and deep veins.

Clinical Notes

The walls of arteries are subject to various disease processes which may lead to hypertrophy or degenerative changes resulting in dilatation or narrowing of the lumen.

An *aneurysm* is a dilatation which may present a fusiform swelling when the whole circumference of the artery is affected; or a saccular swelling when there is weakness in one part of the wall. This tumour-like swelling will press on adjacent structures causing pressure symptoms or it may rupture. An aneurysm of any part of the aorta is very serious. *Arteritis* is inflammation of an artery.

Arteriosclerosis is described as a hardening of the arterial walls, generally associated with hypertension and sometimes with chronic renal disease. Other disorders include *embolus*, a clot moving in the circulation which becomes impacted in a small vessel, and *thrombus*, a clot obstructing a blood vessel at the point where it formed (*see also* page 171).

Atherosclerosis is a progressive disorder often affecting arteries in the lower limbs, causing discoloration, numbness and pain. *Thrombo-angiitis-obliterans* is one form.

Arterial surgery is one of the branches of surgery to advance considerably in recent years. *Arterial grafts* are now available for replacing damaged or diseased segments of vessels (*see* note on page 186).

Veins are subject to *phlebitis*—an inflammation of the vein wall which may be due to infection or injury. *Venous thrombosis*, a blood clot obstructing a vein, or *thrombo-phlebitis*, inflammation of a vein complicated by an obstructing blood clot, may follow.

Varicose veins are dilated and tortuous superficial veins. This condition is attributed to a number of causes and there may be an inherited tendency. Varicose veins most commonly occur in the lower limbs. Varicosities are associated with both the long and short saphenous veins (*see* page 193). Normally the long column of blood in these veins is supported by cusp-like valves at intervals throughout the veins (*see* Fig. 11/2). If these valves degenerate or if the vein lumen widens, the column of blood can drop back, as it were unsupported, and the veins become dilated and tortuous. If the main valves guarding the junction of the superficial with the deep veins become incompetent, blood then flows from the deep into the superficial veins instead of vice versa. This greatly increases the filling and varicosity of the superficial veins.

Haemorrhoids (piles) are dilated veins in the rectum; they are described as *internal* or *external*, depending on the position of the plexus of veins affected. **Oesophageal varices** are rare dilated veins in the oesophagus which may bleed profusely and require treatment.

Chapter 12

THE LYMPHATIC SYSTEM, THE SPLEEN AND THE RETICULO-ENDOTHELIAL SYSTEM

The lymphatic system is intimately connected with the circulatory system. The blood leaves the heart by arteries, and is returned to it by veins. As explained on page 20, some of the fluid which leaves the circulation is returned to it by the lymphatics, which permeate the tissue spaces.

Composition. Lymph is similar to plasma but contains less protein. The lymph glands add lymphocytes to the lymph so that there are numbers of them in the large lymphatic vessels. There are no other cells. The lymph in the vessels is moved onwards by the contraction of the muscles surrounding them aided by the presence of valves in some of the larger lymphatic channels.

Functions. (1) To return fluid and protein from the tissues to the circulation.

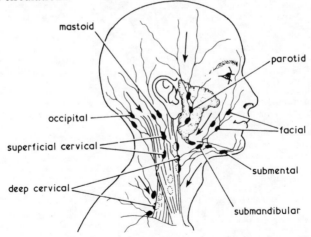

Fig. 12/1.—The Principal Groups of Lymphatic Glands of the Head and Neck

195

(2) To transport lymphocytes from the lymphatic glands to the circulation.

(3) To carry emulsified fat from the intestine to the circulation. The lymphatics performing this function are the lacteals.

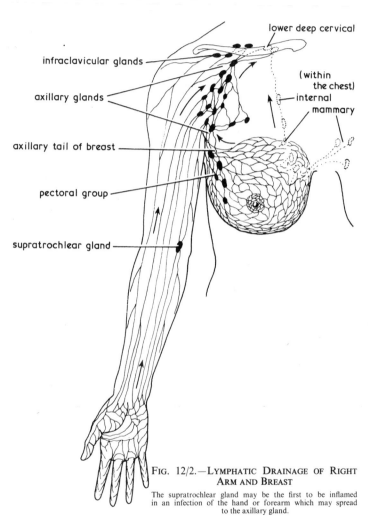

FIG. 12/2.—LYMPHATIC DRAINAGE OF RIGHT ARM AND BREAST

The supratrochlear gland may be the first to be inflamed in an infection of the hand or forearm which may spread to the axillary gland.

(4) The lymphatic glands filter out and destroy micro-organisms in order to prevent infection spreading from the point where the organisms entered the tissues, to other parts of the body.

(5) Following an infection the lymphatic glands produce antibodies to protect the body against subsequent infection.

Lymphatic Vessels. These are similar in structure to the small veins, but have more numerous valves which give the vessels a beaded appearance. The smallest lymphatic vessels or *lymph capillaries* are larger then the blood capillaries, and consist of an endothelial coat only. The lymphatic vessels begin either as minute plexuses of very small capillaries, or in lymphatic spaces, in the substance of the tissues of various organs. Special lymphatic vessels, called *lacteals,* are found in the villi of the small intestine (*see* page 228).

Lymphatic Glands or Nodes are small oval or bean-like bodies, placed in the course of the lymphatic vessels, and joined together by them. They act as filters and are sited where lymphocytes are formed. The main groups lie in the neck, axilla, thorax, abdomen, and groin. (For positions of glands, *see* illustrations.)

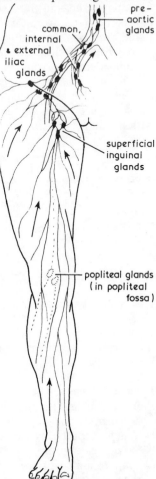

pre-aortic glands

common, internal & external iliac glands

superficial inguinal glands

popliteal glands (in popliteal fossa)

FIG. 12/3.—THE PRINCIPAL LYM-PHATIC GLANDS OF THE RIGHT LOWER LIMB

These glands become inflamed and enlarged in infections of the foot and leg (*see* Clinical Note, page 202).

A lymphatic gland has a convex and a concave border; the latter is called the *hilum*. A gland is composed of fibrous and muscular tissue, and gland substance. An outer fibrous capsule covers the gland; from this, processes of muscular and fibrous tissue, *trabeculae,* pass into the gland, forming partitions; the spaces between these are filled with the gland tissue, which contains numerous white blood cells or *lymphocytes.*

Afferent lymphatic vessels pass through the capsule on the convex border of the gland, and empty their contents into the substance of the gland. This material comes into contact with the numerous lymph corpuscles, and together with many

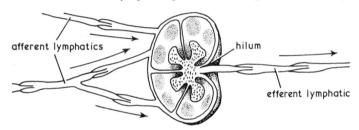

FIG. 12/4.—A LYMPHATIC GLAND, SHOWING AFFERENT VESSELS PASSING TO THE CONVEX BORDER, AND THE EFFERENT VESSEL PASSING FROM THE HILUM OF THE GLAND

When a load of infection is brought by the afferent vessel the gland may fail to deal with it and become inflamed.

of these is collected by *efferent lymphatic vessels* which carry it away from the hilum. The arteries and veins also enter and leave the gland at the hilum.

Lymphatic Ducts. There are two main lymphatic ducts, the thoracic duct and the right lymphatic duct.

The thoracic duct begins as the *receptaculum chyli* or *cisterna chyli* in front of the lumbar vertebrae. It then passes up through the abdomen and thorax, inclining to the left of the vertebral column, to unite with the great veins at the root of the left side of the neck, into which it pours its contents.

The thoracic duct collects the lymph from all parts of the body, except the parts drained by the right lymphatic duct.

The right lymphatic duct, which is a much smaller vessel, collects the lymph from the right side of the head and neck,

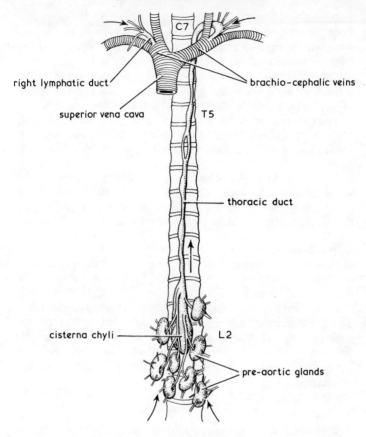

FIG. 12/5.—THE THORACIC DUCT

The lymphatics from the right side of the head and chest and the right arm are seen draining into the junction of the right internal jugular and the right subclavian veins. The thoracic duct terminates by emptying its contents into the same position on the left.

the right upper limb and the right side of the chest, and empties it into the veins at the right side of the root of the neck.

During an infection the lymphatic vessels and glands may become inflamed, as will be seen in the swollen painful glands in the axilla or groin in the case of a septic finger or toe.

The Tonsils also are composed of lymph tissue (*see* page 217). They lie between the pillars of the fauces and are freely supplied with lymphocytes, which lie in the fluid on the surface, and in the crypts of the tonsils.

Considerable quantities of *lymphoid tissue* enter into the formation of the spleen, the serous membranes, and the lining of the small intestine. In the intestine it is contained in the mucous coat; in some parts nodules of lymph tissue are found; when single these are called the *solitary glands,* and when in groups they form *Peyer's patches* (*see* page 228).

The Villi are largely lymph tissue; the central lacteal in the villus communicates with plexuses of lymphatic vessels in the submucous tissue, whence the lymph is passed on, finally reaching the receptaculum chyli (*see* page 233).

Serous Membranes. The serous membranes, the most extensive of which is the peritoneum, will be described in the chapters dealing with the organs with which each is connected. The serous membranes are intimately associated with the lymphatic system. Their various folds carry lymphatics and blood vessels. These membranes are lined with pavement epithelium or *endothelium,* and many small openings are contained in this fine lining. These openings are called *stomata*; they lead into lymphatic vessels, and so prevent lymph collecting in the serous cavities.

THE SPLEEN

The spleen is a dark purplish gland lying on the left side of the abdomen in the left hypochondriac region beneath the ninth, tenth, and eleventh ribs. It lies against the fundus of the stomach and its outer surface is in contact with the diaphragm. It touches the left kidney, the splenic flexure of the colon, and the tail of the pancreas.

The spleen consists of a structural connective tissue framework in the interstices of which is the splenic pulp formed from lymphoid tissue and numerous blood cells. It is covered by a capsule of collagenous and elastic tissue and a few smooth muscle fibres. These latter have little if any functional role in the human spleen. From the capsule processes called trabeculae pass into the substance of the gland breaking it up into compartments.

The splenic blood vessels enter and leave the gland at the hilum, which is on the inner surface. The blood vessels empty their contents directly into the splenic pulp, so that the blood comes into contact with the spleen substance, and is not as in other organs separated from it by blood vessels. There is no ordinary capillary system—the blood comes into direct contact with the cells of the organ. The blood which flows through the spleen is collected in a system of venous sinuses which empty their blood into the branches which unite to form the splenic vein by which the blood is carried from the spleen to enter the portal circulation and be conveyed to the liver.

The Functions of the Spleen. The spleen forms red blood cells during fetal life and may do so in adult life if the function of the bone marrow is impaired.

It separates worn-out red blood cells from the circulation.

The spleen manufactures lymphocytes.

It is also thought to destroy white blood cells and platelets.

As part of the *reticulo-endothelial system* (*see* below) the spleen is concerned in protection from disease; it manufactures antibodies.

The spleen is not essential to life. In some cases of haemolytic anaemia *splenectomy* is performed and as a result of this the fragility of the red cells improves and relief can be obtained.

THE RETICULO-ENDOTHELIAL SYSTEM

Throughout all the tissues and organs of the body there are certain cells which ingest (phagocytose) foreign particles and bacteria. They are particularly concentrated in the lymph glands, spleen, liver and bone marrow. These cells have great powers of multiplication and are related both to lymphocytes and to the blood-forming organs; they are concerned in protecting the body from infection.

Clinical Notes

Lymphangitis is infection and inflammation of the lymphatic vessels when they can be seen as red lines beneath the skin (*see* page 197). *Lymphadenitis,* inflammation of the lymphatic glands which lie on the course of the vessels, is usually secondary to some infection in the field which the group of glands drains, e.g., the swollen glands in the neck in tonsillitis.

Enlargement of the axillary glands occurs in malignant diseases of the breast and in infections of the areas drained by these glands (*see* Fig. 12/2). Radical mastectomy performed for carcinoma of the breast aims at removing all the ramifications of the lymphatic vessels draining this area.

The spleen may be ruptured by direct or indirect violence, or injured by a stabbing incident or gunshot wound. Considerable bleeding occurs owing to the vascularity of the spleen. This may result in shock as the blood in the peritoneal cavity causes irritation. *Splenectomy* may be necessary. See also the note on page 201 on splenectomy in haemolytic anaemia.

When the spleen is enlarged it can be felt below the costal margin. It moves downwards and medially during a deep inspiration.

Leukaemia, which is regarded as a cancerous condition, is characterized by an overproduction of leucocytes. It is classified according to the type of leucocyte affected as lymphatic leukaemia or myeloid leukaemia. The disease may occur at any age, and is commonest in childhood. The condition may be acute or chronic, according to the speed of progress. The *prognosis* is poor; in acute leukaemia death may occur in a few weeks, but patients with chronic leukaemia may live several years.

Chapter 13

THE CLASSIFICATION OF FOOD

Food is needed to build the body, make good wear and tear, and to act as fuel for the production of heat and energy.

Protein is the only class of food which contains nitrogen. Protein is derived from animal and vegetable sources and forms the basic protoplasmic content of every living cell; it is essential for growth, repair and in reproduction.

Examples of Proteins. Proteins may be classified as animal and vegetable proteins, but more accurately they are described as Class A and Class B proteins. A first-class or complete protein food contains at least five of the eight essential amino-acids (altogether there are over twenty).

Class A proteins are all animal proteins.
Myosin in lean meat and fish
Albumin in egg-white, the lact-albumin of milk and the albumin of blood, as contained in lean meat
Caseinogen contained in milk when it is curdled and in cheese
Globulin, such as blood globulin
Vitellin, a substance similar to globulin, found in egg-yolk.

Class B proteins, which are incomplete proteins lacking certain essential amino-acids, are mainly vegetable in origin.

Gluten, the protein of wheat (bread) and other cereals
Legumen (pulses) present in peas, beans and lentils. Soya bean is rich in legumen
Gelatin is an incomplete animal protein derived from animal tissues such as bones and ligamentous tissues. It is present in some meat extracts and in some nutrient broths. It is also obtained from certain vegetable tissues; one example is *agar-agar* used in the preparation of jellies and broths.

Proteins are made up of combinations of *amino-acids*, and in the process of digestion each protein is split up by the enzymes which act upon it into the several amino-acids of which it is formed, as it is only when in this state that protein can be utilized by the tissues of the body. From 80–100 grams of protein are required in a normal diet per day, 50 grams should be class A protein. *Meat, fish, eggs, milk* and *cheese* are rich sources.

Carbohydrates contain carbon which is combined in them with hydrogen and oxygen in the proportions in which these are present in water (H_2O). This class of food supplies the body with heat and energy, the carbon combining with oxygen forming carbon dioxide and producing energy. Carbohydrates yield 17·2 kilojoules (4·1 kilocalories) for every gram used in the tissues. A normal adult eats about 300 grams per day as *sugars* and *starches.*

Sugars may be classed as *monosaccharides* or single sugars, e.g. glucose, fructose, galactose, or as *disaccharides* (double sugars), e.g. sucrose, maltose, lactose.

Natural sugars, with the exception of *lactose,* milk sugar, are derived from the vegetable kingdom:

Cane sugar and beet sugar contain the disaccharide *sucrose.*

Fruits and honey contain *fructose* (also called *laevulose*). Fruits also contain *glucose* (also called *dextrose*).

Maltose, or malt sugar, is a disaccharide formed by the hydrolysis of starch.

Starches are mainly derived from green plant life in the first instance and then stored in the stems, roots and seeds of these plants.

> *Cereals,* wheat (flour), maize and cornflour, barley, rice and sago are examples
>
> *Root vegetables,* particularly matured potatoes, contain considerable starch
>
> *Cellulose* is an example of the type of starch found in the stems and stalks of plants
>
> *Glycogen* is an animal starch present in the muscles and liver

Complex carbohydrates such as starch and cellulose are called *polysaccharides.*

All digested carbohydrates are converted into the simple sugar group, monosaccharides, and are utilized by the tissues of the body in the form of glucose.

Glycogen is the result of the conversion of simple sugars in solution in the body into this animal starch which is the form in which carbohydrates can be stored in the liver and muscles until required, when liver glycogen can be reconverted again into a simple monosaccharide.

Fats are derived from animal and vegetable sources. They are composed of carbon, hydrogen and oxygen and stored as compounds of fatty acids and glycerin.

Examples of *animal fats* are meat and bacon fat and dairy produce such as milk, butter, cheese, and egg-yolk. Animal fats are an essential constituent of diet as they contain stores of Vitamin A and D. Of the *vegetable fats* olive oil and the nut fats are the best-known examples.

Fats are of the same use to the body as carbohydrates, they produce heat and energy. Fat is stored in the body as adipose tissue. It forms the chief reserve store of energy. Fats yield 38·9 kilojoules (9·3 kilocalories) for every gram; a normal adult's daily diet contains 100 grams.

Carbohydrates and fats are the *fuel foods*.

Water forms two-thirds of the weight of the body. It is essential to well-being, and deprivation of water is more immediately serious than of any other article of diet. It forms a large part of the tissues. It dissolves many substances and so helps the chemical changes in the digestive system. It maintains the normal salt concentration of the tissues, thus regulating many of the processes of the body and rendering the process of osmosis possible.

Water is taken in the form of liquids drunk. The amount may be regarded as equal to water. A large proportion of solid food is also composed of water, particularly the fruits and vegetables, which contain from 75 per cent water, as in potato. Many of the fruits contain over 85 per cent water, and melon 95 per cent or more. A certain amount of water in the body is obtained from the oxidation of food.

Water is gained	*Water is lost each day*	
By fluids	As urine	1,500 ml
The water content of food	By the skin	900 ml
Oxidation processes of food	In expired air	400 ml
	In the faeces	200 ml

The water balance of the body must be maintained so that the amount gained is equal to the amount lost. Adjustment is made by increase or decrease in the quantity lost as urine. Thus if a man sweats a great deal, losing more water by the skin, less will be lost as urine. In a disease where larger quantities of urine are voided, as in *diabetes insipidus* and *diabetes mellitus*, thirst is experienced and more fluid is taken by mouth.

Any excess of water gained (over loss) results in *oedema*; conversely, any excess of water lost (as for example in diarrhoea and/or vomiting) over fluid intake, would result in *dehydration*. (*See also* Balance of Fluid in the Body, page 20.)

Oxygen is essential to the body in the oxidation of foodstuffs. Moreover oxygen is needed for nearly all the metabolic activities going on in the body and for its survival.

Salts. There are various salts in the body which form ·the mineral content of most foods.

Calcium is supplied by milk, cheese, egg-yolk, and by many vegetables, particularly cabbage and carrots. It is required by all tissues, is carried by the blood serum and its use is regulated by the parathyroid secretion. It is particularly necessary for the ossification of bone, the formation of the teeth, and the clotting of blood.

Sulphur is supplied by all protein substances. It is essential for the well-being of all tissue.

Iron is present in meat, eggs, cheese, bread, and green vegetables. It is needed for the composition of haemoglobin, and in combination with it oxygen is distributed to the body. A deficiency of iron leads to iron-deficiency anaemia (*see* clinical note, page 176).

Sodium chloride is present in most food and also supplied

as table salt. It is the most abundant salt in the *extracellular tissue fluids*.

·*Potassium* is present in nearly all food, particularly protein-containing substances. And it is the most abundant salt in the *intracellular tissue fluid*.

Phosphorus is present in every cell in the body. It is essential for the production of muscular and nervous energy and for the correct composition of hard tissues such as bone and dentine. Phosphorus is supplied in milk, egg-yolk, fish-roe, and green vegetables.

Iodine is present in the products of the sea and is found in foods grown near the sea. The presence of it in the body balances the metabolic processes stimulated by the secretion of the thyroid gland. It may be deficient in some inland areas, when it is then provided as iodized table salt or sweets, to which potassium iodide has been added.

Vitamins are compounds which are essential to life, health, and growth; they are concerned with the well-being of body metabolism. These compounds, classified according to their solubility as *fat-soluble* and *water-soluble* vitamins, are normally absorbed in the small intestine.

Fat-soluble Vitamins include A, D, E and K. *Vitamin A* is found in fish-liver oils, milk and dairy produce, liver and certain fatty fish. It is *growth-promoting* and *anti-infective*.

Carotene, a precursor of Vitamin A, is found in leafy vegetables, carrots and in some fruits. It is converted to A in the body.

Deficiency leads to epithelial degeneration and predisposes to infection, causes *night blindness* and a disease of the eyes known as *xerophthalmia*.

Vitamin D (D_2 *Calciferol*) is produced by the action of ultra-violet light on Ergosterol (irradiated ergosterol), and D_3, the *natural Vitamin D*, is produced in the body by exposure of the skin to sunlight or ultraviolet light. These vitamins are found in fish-liver oils, eggs, butter and fatty fish.

D is essential for the growth of bone and teeth as it promotes calcium absorption. It is the *anti-rachitic vitamin*.

Deficiency causes rickets in infants and children, and osteo-malacia in adults.

Vitamin E is found in wheat germ oil, egg-yolk, milk, and some green vegetables. It is described as the *anti-sterility vitamin*, but little is known of this function in man.

Vitamin K is found in alfalfa, spinach, soya bean and pig's liver. It is required for the formation of prothrombin (*see* page 171). *Insufficiency* results in prolonged clotting time in some haemorrhagic diseases of the newly born and when there is bile in the blood as in obstructive jaundice.

Water-soluble Vitamins include Vitamin B_1, the B-complex, Vitamins C and P.

Vitamin B-Complex contains the following, and all are essential to good health:

Aneurin or B_1 (Thiamine) is found in whole grain, pulses, pig's liver, eggs and yeast. It is the *anti-beri-beri* vitamin, essential for the metabolism of carbohydrate.

Riboflavin or B_2 is found in wheat germ, milk, liver, soya bean, peas and lentils. *Deficiency* causes dermatitis and cracks and fissures about the lips and nose.

Nicotinic acid (Niacin) is found in lean meat, liver, wheat germ, green vegetables. It is the P.P. factor or *anti-pellagra* vitamin.

Biotin (formerly called Vitamin H) is found in liver, kidneys, mushrooms, milk, eggs, yeast and nuts. It is essential for the health of the skin and mucous membrane. *Deficiency* gives rise to dermatitis and conjunctivitis.

Pyridoxine (*Vitamin B_6*) is found in green leaves, yeast, liver and kidney. It is concerned with the formation of normal red blood cells.

Vitamin B_{12} is found in liver. It is used in the treatment of pernicious anaemia.

Other vitamins of the B-complex include *Choline, Pantothenic Acid, Folic Acid*, and *Para-amino Benzoic Acid*.

Vitamin C is the *anti-scorbutic* vitamin. It is found in many fruits, particularly the citrus fruits and blackcurrants, rose hips and in vegetables. It is essential for the healthy development of all connective tissues. It raises immunity to infection and assists in the healing of wounds and fractures. *Deficiency* causes scurvy.

Vitamin P (Hesperidin) is found in citrus fruits, blackcurrants, rose hips and green leaves. It aids in maintaining normal capillary resistance. *Insufficiency* may result in subcutaneous bleedings.

Clinical Notes

Nutrition in relation to health. The World Health Organisation defines health as 'a state of complete physical, mental and social well-being and not merely absence of disease or infirmity'. The starving peoples in famine countries are the unfortunate examples of severe *malnutrition*; **kwashiorkor,** which is a serious disease, particularly in infants, is due to lack of protein.

An adequate balanced diet is essential to individuals of all ages. The value of *antenatal care* by balancing the diet of the pregnant mother is seen in our 'bonny babies'.

Environment and diet should be considered together. Nutritional requirements include adequate proportions of protein, fat, carbohydrate, water, minerals and vitamins as outlined in the previous pages. Pregnant women, infants and young children, school children and adolescents, families in the lower income groups and elderly people living on a retirement pension, may be vulnerable to malnutrition because, for one reason or another, they do not get a well-balanced diet. Some need help to balance their budget and in purchasing and preparing food; others may require financial assistance. A happy environment in a comfortable home with friends and interesting occupations are also essentials which can help in the enjoyment and proper use of available food.

Under-nutrition is self-explanatory; *starvation* is an extreme degree. About half the world's population is not adequately nourished, and one-third is definitely suffering from food shortages. Millions of children are undersized, emaciated and apathetic, a number are starving, which results in so low a resistance to disease that when not dying of actual starvation they are vulnerable to numerous infections. A great proportion do not survive beyond infancy and early childhood.

Over-nutrition is a problem in the affluent countries and amongst the more well-to-do in all countries. It is not that people necessarily eat more, but the dishes put before them, often in two main meals a day have higher food value, and consist of rich meats with their accessories, cream, succulent cheese, additional butter and eggs, often accompanied by wines and spirits. This diet increases the tendency to certain forms of heart and arterial disease, to obesity with its train of disabilities and to diabetes.

Inadequate diet in illness. There are many reasons why a patient may be unable to take enough food by mouth, to mention a few:

Nausea, loss of appetite and vomiting, due for example to disease of the stomach, or to uraemia.

Indifference to food may be met in many physical conditions (*see* below) and in some emotional states, as in fear, grief and despair.

Pain renders swallowing difficult, as in tonsillitis, quinsy, after tonsillectomy, in fractures of the face and jaw.

Those *breathless,* as in asthma and bronchitis, may be unable to chew; when breathing is difficult swallowing is inhibited.

Weakness of the muscles of mastication may arise in facial paralysis (*see* page 343) and in certain neurological conditions.

A patient who has had a *stroke* may find chewing and swallowing difficult or impossible.

A *persistently drowsy patient* or one in *stupor* may be roused to take food but is very unlikely to get enough; one who is *unconscious* cannot swallow.

Tube Feeding. Whenever a patient is unable to take sufficient food, feeding by a naso-gastric tube is instituted, either to supplement or entirely supply his dietary requirements, by means of:

A *liquid diet* of Complan and glucose so constituted that it contains all the essential elements, including sodium, potassium, and vitamins.

A *liquidized well-balanced diet* may alternatively be given.

It cannot be too strongly emphasized that *a patient inadequately fed is being starved*; this is detrimental to his recovery and well-being, demanding immediate steps to correct the situation.

METABOLIC RESPONSE IN ILLNESS

A metabolic response occurs in illness and injury and after major surgery which is characterized by increase of protein breakdown and muscle wasting. In a short acute illness these changes take place so rapidly that considerable damage results which may seriously retard recovery. In chronic illnesses the process is more gradual. *Treatment* is urgently needed by tube-feeding when necessary, either to supplement or provide a diet commensurate with the medical condition of the patient which will combat these metabolic changes (*see* above note).

There is some degree of metabolic disturbance after any illness, however simple, and also after trauma and surgery. A young athlete, after a meniscectomy, for example, may not feel quite himself for 24 to 36 hours, having headache, malaise, diminished urinary output and a tendency to constipation.

A woman of 40, having recently had hysterectomy, may feel distinctly unwell for 3 to 4 days, describing her state as similar to being a victim in a road accident. This is a fair analogy, the difference being that under surgery a general anaesthetic is given and the patient is not conscious of pain, but the tissues have been injured nevertheless.

In a road accident, if conscious, the victim experiences fear and is aware of every painful impression resulting from the injuries sustained.

It is important, therefore, to make detailed accurate observations of the general condition of a patient, having regard to his state of health, age, mental and emotional condition, and to obtain medical advice, whenever in doubt, before applying as a routine measure for example the rules for early ambulation.

Chapter 14

THE ALIMENTARY CANAL AND THE DIGESTION OF FOOD

The digestive system deals with the reception of food and with the preparation of it for assimilation by the body. The *alimentary canal* consists of the following parts:

Mouth
Pharynx
Oesophagus
Stomach
Small and Large Intestine.

In addition the mouth contains the *teeth*, which masticate the food, and the tongue which assists in taste and swallowing. Several glands or groups of glands pour important *digestive fluids* into the alimentary tract:

The *Salivary Glands*, whose ducts enter the mouth,
The *Pancreas* and *Liver*, described in Chapter 15.

The entire alimentary canal is lined by mucous membrane; from the lips to the end of the oesophagus this is stratified epithelium. From the stomach to the anal canal it is composed of columnar cells, and in the anal canal of stratified epithelium.

During the processes of digestion food is broken down into simple substances which can be absorbed and used by the cells of the body tissues. The various changes in the character of food are brought about by the activity of ferments or *enzymes* contained in the different digestive fluids. These substances have a specific action—they select and act on one type of food and have no effect on other types.

Ptyalin (salivary amylase) for example acts only on sugars and starch and *pepsin* only on protein. One digestive fluid, the pancreatic fluid for example, may contain several enzymes, each enzyme acting only on one type of food.

An Enzyme is a chemical substance which produces changes in the chemistry of other substances, without itself undergoing any change. The healthy action of the various enzymes depends to a great extent on the presence of mineral salts, and the correct acidity or alkalinity.

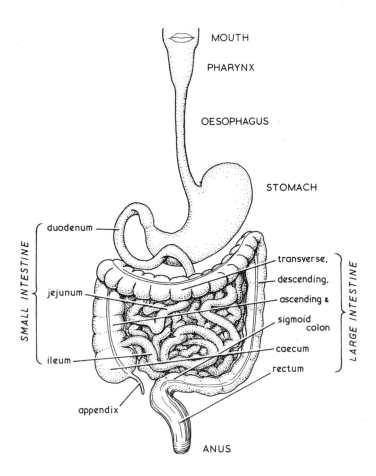

Fig. 14/1.—An Outline of the Alimentary Tract

THE MOUTH

The mouth is an oval cavity at the beginning of the alimentary canal. It consists of two parts: an outer small part, the *vestibule*. which is the space between the gums and teeth, and the lips and cheeks; and the inner part, the *cavity of the mouth*, which is bounded at the sides by the maxillary bones and the teeth, and communicates behind with the oral pharynx. The *roof of the mouth* is formed by the palate, and the tongue lies in the *floor* attached to the hyoid bone. In the middle line a fold of mucous membrane (*the frenulum linguae*) connects the tongue

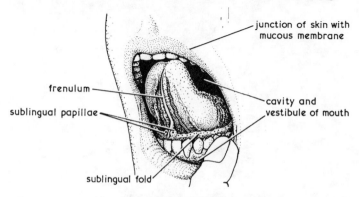

junction of skin with
mucous membrane

frenulum

sublingual papillae

cavity and
vestibule of mouth

sublingual fold

FIG. 14/2.—THE MOUTH

with the floor of the mouth. On each side of this lies the *sublingual papilla* which contains the opening of the submandibular salivary gland; slightly external to this papilla lies the *sublingual fold*, where the tiny openings of the sublingual salivary gland lie.

The mucous membrane of the mouth is covered by stratified epithelium. Beneath this lie tiny glands, which secrete mucus. This membrane is very vascular, it also contains numerous sensory nerve endings.

The lips are two fleshy folds which form the orifice of the mouth. They are covered on the outer surface with skin and on the inner surface with mucous membrane. The orbicularis oris muscle closes the lips; the levator anguli oris raises, and the depressor anguli oris depresses, the corners of the mouth. The junction of the upper and lower lips form the angle of the mouth.

The palate consists of two parts, *the hard palate*, which is composed of the palatine processes of the maxillary bones in front and the palatine bones farther back; behind this lies the *soft palate*, which is a movable hinged flap of fibrous tissue and mucous membrane. Its movements are controlled by its own muscles. From the middle of the soft palate a conical process,

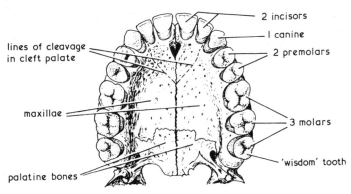

FIG. 14/3.—THE UPPER TEETH AND BONY PALATE

the *uvula*, hangs. Arching downwards and outwards from this are the *pillars of the fauces,* between which are double folds of muscle and mucous membrane in which lie the tonsils (*see* Fig. 24/1, page 367).

The cheeks form the fleshy sides of the face and are joined to the lips at the naso-labial fold which runs from the side of the nose to the corner of the mouth. The cheeks are lined by mucous membrane which contains tiny papillae. The muscle of the cheek is the buccinator.

The Teeth and Mastication. There are two sets of teeth, *the temporary set* and *the permanent set*. There are twenty temporary or milk teeth, ten in each jaw, named from the middle line on each side, two incisors, one canine, two molars. The permanent teeth are increased to thirty-two, sixteen in each jaw, as follows: named from the centre, two incisors, one canine, two pre-molars, three molars.

As a rule an infant cuts his first tooth at the age of six months. The central incisors of the lower jaw are cut first, then the lateral incisors; the first molars at about twelve to fifteen months, the canines at eighteen months, and lastly at about twenty months the remaining molars.

An infant of twelve months should have eight teeth, two central and two lateral incisors in upper and lower jaws. At the age of two the child has the complete temporary set of teeth. The teeth in the lower jaw are cut before the corresponding teeth in the upper jaw as a rule.

The *permanent teeth* begin to replace the temporary ones at about the age of six years. A molar is cut first behind the tem-

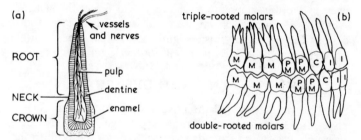

FIG. 14/4.—(a) SECTION OF TOOTH (b) PERMANENT TEETH OF THE RIGHT SIDE OF THE JAWS

porary teeth on each side, then the incisors at seven to eight years, pre-molars nine to ten years, canines at eleven years, second molars about twelve years, and the last molars which are called 'wisdom teeth' later.

A tooth possesses a *crown*, a *neck*, and a *root*. The crown projects above the gum, the neck is surrounded by the gum and the root lies beneath it. A tooth is made of a very hard material, *dentine*; in the centre of the structure is the *pulp cavity*. Tooth pulp contains connective tissue cells, blood vessels, and nerves. The part of the tooth projecting above the gum is covered with *enamel*, which is much harder than dentine.

Mastication is the biting and grinding of food between the upper and lower teeth. Movements of the tongue and cheeks assist, by manipulating the soft foods against the hard palate and the teeth.

The chief *muscles of mastication* are the masseter, temporalis, and the medial and lateral pterygoid muscles.

The tongue is described on page 366.

Dental hygiene should be stressed. Young children can learn to brush their teeth up and down, inside and out, after all meals and before going to bed. Any dental paste or powder may be used. Snacks and sweets should not be eaten between meals, or in bed; this is a common source of dental trouble.

The *eruption of the teeth*, both temporary and permanent, should be supervised; regular visits to the dentist are essential every 3 months, if possible, or every 4 to 6 months at least. Absence of pain does not mean absence of disease, tooth caries. Teenagers may visit slightly less often, then in adult life visits may be less frequent but not less regular.

THE PHARYNX AND OESOPHAGUS

The Pharynx lies behind the nose, mouth, and larynx. It is a cone-shaped musculo-membranous passage with the widest part uppermost and extends from the base of the skull to the level of the sixth cervical vertebra, that is, the level of the cricoid cartilage where the pharynx joins the oesophagus. N.B. At this level also the larynx joins the trachea. The pharynx is about 12·5 cm (5 inches) long and is divided into three portions:

The nasopharynx, behind the nose; into the walls of this part the Eustachian tubes open. The adenoids lie in the nasopharynx.

The oral pharynx, behind the mouth; the tonsils lie in the lateral walls of this part of the pharynx (*see* below).

The laryngeal pharynx, which is the lowest part, and lies behind the larynx.

There are seven openings into the pharynx—two Eustachian tubes in the walls of the nasopharynx, two posterior nares from behind the nasal cavities, and the mouth, larynx and oesophagus (*see* page 253).

Structure of the Pharynx. The pharynx is composed of three coats, mucous, fibrous, and muscular. The inner mucous coat is continuous with the lining of the nose, mouth and Eustachian tubes; in the upper part of the pharynx this is respiratory epithelium, continuous with that of the nose. The lower part of the pharynx, continuous with the mouth, is lined with stratified epithelium. The fibrous coat lies between the mucous and muscular coats. The chief muscles of the pharynx are the *constrictor muscles*, which contract on the food received into the pharynx, and force it on to the oesophagus.

The tonsils are two collections of lymphoid tissue placed one on each side of the pharynx between the pillars of the fauces. They are permeated with blood and lymphatic vessels and contain masses of lymphocytes. The surface of the tonsil is covered with mucous membrane continuous with that of the lower part of the pharynx. This surface is studded by crypts and into these crypts numerous mucus-secreting glands pour their secretion. This mucus contains many lymphocytes. In this way, the tonsils act as the first line of defence in infection spreading from the nose, mouth and throat; nevertheless they may fail to resist the infection when *tonsillitis* (inflammation) or *quinsy*, a peritonsillar abscess, may arise. After treatment with antibiotics and local treatment, *tonsillectomy* may be considered, but less often today than formerly.

The mucous membrane of the pharynx near the opening of the posterior nares and the Eustachian tubes also contains lymphoid tissue rather like that of the tonsils. When this tissue is hypertrophied it obstructs the posterior nares and the condition described as *enlarged adenoids* is produced.

The Oesophagus is a muscular tube 23 to 25 cm (9 to 10 inches) long, reaching from the pharynx above, to the cardiac orifice of the stomach below (*see* Fig. 14/1, page 212). It lies behind the trachea to which it adapts and in front of the vertebral column. Passing through the thorax it pierces the diaphragm, to enter the abdomen where it communicates with the stomach.

The oesophagus consists of four coats, an outer loose connective tissue layer, a *muscular* coat, composed of two layers of muscle fibres, longitudinal and circular, a *sub-mucous* coat and an inner *mucous* membrane.

Swallowing. The act of swallowing follows mastication and may be described in three parts: A voluntary act in which the food is formed into a *bolus*, by the action of the tongue and cheeks, and passed from the back of the mouth into the pharynx.

When the food enters the pharynx, the soft palate rises to shut off the posterior nares, the glottis closes by contraction of its muscles and the constrictor muscles of the pharynx grasp the food and pass it on to the oesophagus; at this moment breathing ceases or choking would occur. One cannot swallow and breathe at the same time. This part of swallowing is a reflex action.

The food passes through the oesophagus by *peristaltic action,* the circular muscle fibres relax in front of the food and contract behind it, and peristaltic waves convey the bolus of food to the stomach.

The second and third parts of swallowing are involuntary, and the first part, although a voluntary act, is for the most part performed automatically.

The oesophagus may be affected by *cardiospasm* or *achalasia* due to failure of the motor function, with absence of peristalsis in the lower part of the oesophagus and failure of the cardiac sphincter to relax. The principal symptoms are dysphagia and regurgitation. *Conservative treatment* by eating easily swallowed foods slowly may help; or *dilatation of the cardiac sphincter* may be undertaken. When failing relief by these means, surgical intervention will be considered.

THE SALIVARY GLANDS AND SALIVA

The salivary glands are compound racemose glands which means they are composed of groups of sac-like alveoli, which form small lobules; ducts from each alveolus unite to form a larger duct which conveys the secretion towards a main duct through which the salivary secretion is poured into the mouth.

The principal salivary glands are the *parotid*, *submandibular*, and *sublingual* glands.

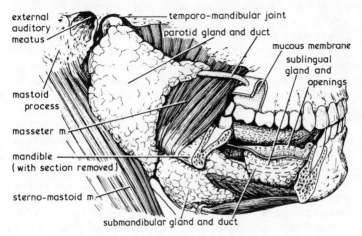

FIG. 14/5.—THE SALIVARY GLANDS

The Parotid Glands are the largest. These lie one on each side, below and slightly in front of the ear. They pour their secretion into the mouth through the parotid or *Stensen's duct*, which opens on the inside of the cheek, opposite the second upper molar tooth. The parotid gland is traversed by two important structures, the external carotid artery and the 7th cranial (facial) nerve.

The Submandibular Glands are the next largest. These lie on each side beneath the jaw-bone, and are about the size of a walnut. Their secretion is poured into the mouth through the submandibular or *Wharton's duct*, which opens into the floor of the mouth, near the *frenulum linguae* (*see* page 366).

The Sublingual Glands are the smallest pair. These lie beneath the tongue on each side of the *frenulum linguae*, and pour their secretion into the floor of the mouth, through several small openings.

The function of the salivary glands is the secretion of saliva which is the first digestive fluid to act upon the food. The flow of saliva is stimulated by (*a*) the presence of food in the mouth; (*b*) the sight, smell and thought of food.

Any of the salivary glands may become infected, but principally the parotids owing to their proximity to the mouth, and also to the possibility of obstruction of the parotid duct. The condition is one of *parotitis* or *parotiditis*, but acute parotitis is rare.

Mumps is infective or epidemic parotitis.

Saliva is a watery, alkaline fluid. It contains a very small proportion of solids, mucin, and a starch-splitting ferment *ptyalin*.

Function. The action of saliva is both physical and chemical. *By its physical action* it moistens the mouth, cleanses the tongue and makes speech easier. It lubricates the food in the mouth and makes swallowing easier, and by moistening the food it dissolves particles, so that the chemical action upon these is facilitated.

The Chemical action of Saliva is due to a ferment *ptyalin* (salivary amylase), which in an alkaline medium acts on sugar and cooked starches. Ptyalin can only act on starch when the cellulose covering of the starch granules has been burst as by cooking, and then the cooked starches are converted into a soluble form of sugar, maltose. This action commences in the mouth, the saliva is swallowed with food and the action of ptyalin continues in the stomach for about twenty minutes or until the food is rendered acid by the action of the gastric fluid.

THE ABDOMINAL CAVITY

The abdomen is the largest cavity in the body. It is oval in shape and extends from the diaphragm above to the pelvis below. The abdominal cavity is described in two parts—the abdomen proper, which is the upper and larger cavity, and the pelvis, the lower and smaller cavity.

Boundaries of the Abdomen. *Above*, the diaphragm. *Below*, the brim of the true pelvis. *At the front and sides*, the abdominal muscles, the iliac bones, and the lower ribs. *At the back*, the vertebral column, and the psoas and quadratus lumborum muscles.

Contents of the Abdomen (*see* Fig. 2/6, page 53). The greater part of the alimentary canal, i.e. the *stomach*, and *small and large intestines*.

The *liver* occupies the right upper part, lying beneath the diaphragm, and overlapping the stomach and the first part of the small intestine. The *gall-bladder* lies beneath the liver.

The *pancreas* lies behind the stomach, and the *spleen* lies near the tail of the pancreas.

The *kidneys* and *adrenal* (*suprarenal*) *glands* lie on the posterior abdominal wall. The *ureters* pass through the abdomen from the kidneys.

The *abdominal aorta*, the *inferior vena cava*, the *receptaculum chyli* and part of the *thoracic duct* lie in the abdomen. *Lymphatic vessels* and *glands*, *nerves*, the *peritoneum* and *fat* are also contained in this cavity.

THE STOMACH AND GASTRIC DIGESTION

The stomach is the most dilatable portion of the alimentary canal. It lies mainly in the epigastric region of the abdomen, and partly in the left hypochondriac and umbilical regions (*see* Fig. 2/4, page 50). It consists of an upper part, the *fundus*, the main body, and a lower horizontal part, the *pyloric antrum*. It communicates with the oesophagus by means of the cardiac orifice or *cardia*, and with the duodenum by the pyloric orifice.

The stomach lies below the diaphragm, in front of the pancreas, and the spleen lies against the left side of the fundus.

Structure. The stomach consists of four coats:

An outer peritoneal coat, which is a serous covering.

A muscular coat, which is in three layers, (*a*) *longitudinal fibres*, which lie superficially and are continuous with the muscle of the oesophagus, (*b*) *circular fibres*, which are thickest at the pylorus where they form the sphincter muscle, and lie beneath the first layer, and (*c*) *oblique fibres*, which are found chiefly at the fundus of the stomach and pass from the cardiac orifice and sweep downwards over the lesser curvature.

A sub-mucous coat of areolar tissue contains the blood vessels and lymphatics.

A mucous coat, the inner membrane, is thick and soft, and is arranged in corrugated folds, rugae, which disappear when the organ is distended by food.

The mucous membrane is lined by columnar epithelium and it contains numerous lymphatics. All the cells secrete mucus. The surface is covered by the tiny ducts of the gastric glands. These

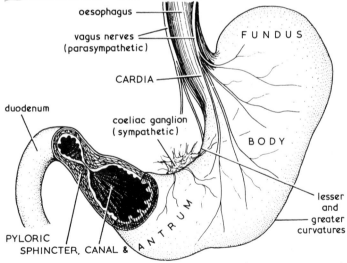

FIG. 14/6. —The Anterior Aspect of the Stomach (a portion opened to show the lining)

lead from the branched tubular gastric glands, and the ducts opening on to the surface are lined by columnar epithelium continuous with that of the mucous surface of the stomach. The epithelium of the secreting part of the gland is modified and varies in different areas of the stomach.

Cardiac glands lie nearest to the oesophageal opening. These are tubular glands, either simple or branched and secrete an alkaline mucus.

Glands of the fundus predominate; they are tubular glands and contain different types of cells. Some cells produce *pepsin*

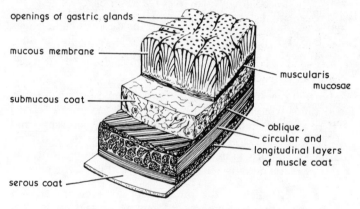

openings of gastric glands

mucous membrane

muscularis mucosae

submucous coat

oblique, circular and longitudinal layers of muscle coat

serous coat

FIG. 14/7.—STRUCTURE OF STOMACH WALL

—the peptic cells; some produce the *acid* contained in the gastric juice—acid or oxyntic cells; and others produce *mucin*.

Pyloric glands. The glands in the pyloric canal are also tubular in character. They produce mainly alkaline mucus.

Blood and nerve supply. The stomach receives a very liberal blood supply from the gastric and splenic arteries; the nerve supply is derived from the vagus and from the coeliac plexus of the sympathetic system.

Function. The stomach receives the food from the oesophagus through the cardiac orifice, and acts as temporary storage whilst muscular contractions mix the food with the gastric juice. Peristaltic waves begin high up on the fundus, following on each other about three times a minute and passing gently towards the pylorus.

The passage of the food into the stomach during a meal is practically continuous; but the passage of food out of the stomach does not begin at once. Food must first be rendered liquid, then small quantities, about 14 g (half an ounce) at a time, are passed through the pyloric opening into the duodenum. The stomach contents are very acid, and the contents of the duodenum are less acid, and when a small quantity of the acid stomach contents enters the duodenum the pyloric sphincter closes until the acid contents are partly neutralized by the action of the alkaline juices of the duodenum, pancreas and bile. When the sphincter muscle relaxes again the duodenum receives another instalment of stomach contents.

The glands in the mucous coat of the stomach secrete an important digestive fluid, *gastric juice*. This is clear colourless acid fluid. It contains 0·4 per cent of free *hydrochloric acid* (HCl), which acidifies all foods and acts as an antiseptic and disinfectant, rendering many organisms taken in with food harmless, and providing a medium for the digestion of protein foods. Several digestive enzymes are present in gastric juice:

Pepsin, which in the presence of hydrochloric acid is obtained from pepsinogen and acts on protein foods, converting these into more soluble substances called *peptones*;

Rennin, a milk-curdling ferment, which forms *casein* from the soluble caseinogen. Casein is milk protein, and thus separated it can be acted upon by the ferment pepsin. ('Rennet', which is rennin extracted from the stomachs of calves, is available to make junket or to curdle milk in making whey.)

A fat-splitting ferment described as *gastric lipase*—in order to distinguish it from the lipase of the pancreatic juice —is present in small amounts in the stomach, and the digestion of fats commences here.

The stimulation of the secretion of gastric juice is partly nervous and partly chemical. Secretion starts at the very beginning of a meal when the sight and smell of food stimulates the secretion. This is often called the 'physical' phase. The taste of food then stimulates a further nervous secretion. Food in the stomach produces chemical stimulation by causing the wall of the stomach to liberate the hormone (chemical excitor) called *gastrin.*

The secretion of gastric juice may be inhibited by the sympathetic nervous system, as may happen in strong emotion such as anger or fear. We speak of a person being sick with fear, and in this case the stomach may actually reject its contents.

Normal gastric fluid also contains a ferment known as the *blood-forming factor of Castle.* This factor is necessary for the absorption of vitamin B_{12}, *cyanocobalamin* (the haematinic principle). Its absence causes pernicious anaemia.

Summary of the Functions of the Stomach

(1) The stomach receives the food and acts as a reservoir for a short time.

(2) All foods are liquefied and mixed with hydrochloric acid, and in this way prepared for intestinal digestion.

(3) Proteins are converted into peptones.

(4) Milk is curdled and casein set free.

(5) The digestion of fat commences in the stomach.

(6) An anti-anaemic factor is formed.

(7) Chyme, that is liquefied stomach contents, is passed on into the duodenum.

Clinically the movements and conditions of the stomach may be examined by X-rays; by passing a gastroscope for direct viewing; by gastric photography; the normality of gastric secretion may be examined by one of the test meals available. (*See* further Clinical Notes, p. 238.)

THE SMALL INTESTINE AND INTESTINAL DIGESTION

The small intestine is a tube which is probably about 2·4 m (8 feet) long in life. The usual figure given of 6 m (20 feet) is a postmortem finding when the muscle has lost its tone. It extends from the stomach to the ileo-colic valve where it joins the large intestine.

The small intestine lies in the umbilical region of the abdomen and is surrounded by the large intestine. It is divided into several parts (*see* Figs. 14/1 and 14/9).

The Duodenum, the first 25 cm (10 inches) of the small intestine, is shaped like a horse-shoe, the curve encircling the head of the pancreas. The bile and pancreatic ducts open into the duodenum at an orifice known as the *hepatopancreatic ampulla* of

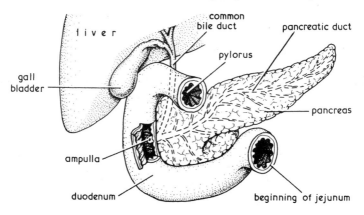

Fig. 14/8.—The Duodenum showing the Opening of the Bile and Pancreatic Ducts at the Hepatopancreatic Ampulla (of Vater)

the bile duct, or the *ampulla of Vater*, 10 cm (4 inches) from the pylorus.

The Jejunum constitutes the upper two-fifths of the remaining small intestine.

The Ileum constitutes the last three-fifths.

Structure. The small intestine is composed of the same four coats as the stomach.

The *outer coat* is a serous membrane, the peritoneum, which closely invests the intestine.

The *muscular coat* consists of two layers of fibres only, an outer layer of *longitudinal fibres*, and beneath these an inner thick layer of *circular fibres*. Between these layers of muscular fibres lie blood vessels, lymphatics, and a plexus of nerves.

A *sub-mucous* coat lies between the circular muscle and the innermost coat or lining. This sub-mucous coat is composed of areolar tissue; it contains numerous blood vessels, lymphatics, glands, and a nerve plexus called the plexus of Meissner. In the duodenum there are some characteristic glands known as *Brunner's glands*. These are tiny racemose glands which secrete a viscid alkaline fluid, which serves to protect the lining of the duodenum from the action of the acid gastric contents.

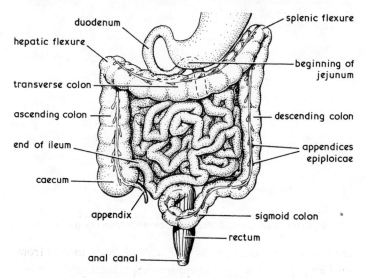

FIG. 14/9.—THE SMALL AND LARGE INTESTINE

Separating the sub-mucous and mucous coats is a layer of plain muscle called the *muscularis mucosae*. Fibres from this pass up to the villi and by their contractions serve to aid in emptying the lacteals (*see* Fig. 14/11).

The *inner mucous* lining is arranged in permanent tuck-like folds, called *valvulae conniventes*, which give it the appearance of fine pleating. These folds increase the extent of the secreting and absorbing surface. They also tend to prevent the too rapid passage of the contents along the intestine, thus giving the digestive juices longer time to act on the food. The mucous coat contains the *crypts of Lieberkühn* which open on to the

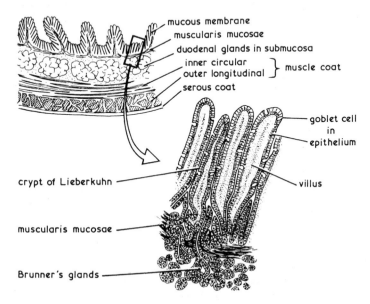

FIG. 14/10.—MICROSCOPIC APPEARANCE OF A SECTION OF THE COATS OF THE DUODENUM

surface between the villi. These are simple tubular glands (*see* Fig. 14/10) lined by columnar epithelium. This epithelium is continuous with that covering the villi.

Several varieties of cells including many leucocytes lie in the mucous coat. Here and there nodules of lymphatic tissue are found, these are called the solitary glands. In the ileum masses of these nodules are present. They constitute the *Peyer's patches* and may contain 20 to 30 solitary glands and measure from one to several centimetres long. These glands exercise a protective function and are *the site of inflammation* in enteric (typhoid) fever.

Glands of the small intestine

Name	Character	Position	Function
Crypts of Lieberkühn	Simple tubular glands	Throughout the mucous membrane of the small intestine	Probably secrete intestinal juice, succus entericus
Brunner's glands	Small racemose glands	Sub-mucous coat of the intestine, particularly the duodenum	Secretion of alkaline substance protective to the duodenum
Solitary glands	Single follicles or nodules of lymphatic tissue	Throughout the mucous membrane of the small intestine	Protection of the intestine from bacterial invasion
Peyer's glands	Groups of solitary glands	Mucous surface of ileum	

To the naked eye the surface of the *valvulae conniventes* has the appearance of soft velvet due to the presence of minute hair-like projections called *villi*.

The Functions of the Small Intestine are digestion and absorption of the *chyme* from the stomach. The contents of the duodenum are alkaline.

The fluid contents (or *chyme*) are passed along the small intestine by a series of rapid *peristaltic movements* lasting a second, with a rest of a few seconds in between. Two other movements are also described.

A *segmental movement*, in which segments of the intestine are cut off by constricting movements of the circular muscle fibres. This enables the liquid contents to be retained temporarily in contact with the intestinal wall, for digestion and absorption. The segments disappear to reappear farther along the organ.

A *pendulum or swaying movement* which causes a mixing together of the intestinal contents.

Two important digestive fluids are passed into the duodenum

by ducts, the *bile* from the liver and the *pancreatic juice* from the pancreas.

Bile is necessary for the digestion of fat which it emulsifies (i.e. breaks up into small particles) and thus helps the action of lipase. It is alkaline and helps to neutralize the acidity of the food leaving the stomach.

Bile salts reduce the surface tension of the intestinal contents and aid in forming an emulsion of the digested fat.

Pancreatic Juice contains three digestive enzymes which act respectively on all three classes of food. It is alkaline.

Amylase digests carbohydrates; it is more powerful than ptyalin, and acts on uncooked as well as on cooked starches converting them into disaccharides.

Lipase is a fat-splitting enzyme, breaking fats up into glycerin and fatty acids. It is most powerful when acting in conjunction with bile.

Trypsin digests proteins. It is produced by the enzyme trypsinogen, present in the pancreatic juice, and converted into the digestive ferment trypsin, by one of the enzymes of the succus entericus, *enterokinase*. Trypsin is more powerful in action than the enzyme pepsin of the gastric juice. It reduces proteins and peptones to the polypeptide group.

A milk-curdling enzyme is also thought by some physiologists to be present in the pancreatic juice.

Succus Entericus. Several enzymes are present in the succus entericus or intestinal juice which complete the digestion of all foods.

Enterokinase activates the proteolytic enzyme of pancreatic juice as described above.

Erepsin completes the digestion of already altered proteins, converting polypeptides into the various amino-acids.

Three enzymes act on carbohydrates, completing the digestion of starches by converting disaccharides into monosaccharides.

Invertase (*sucrase*) acts on cane sugar.

Lactase splits lactose into glucose and galactose, which is then converted into glucose in the liver.

Maltase converts maltose into dextrose.

By the action of the various digestive juices, saliva, gastric juice, pancreatic juice, and the succus entericus, the different food materials have by now been reduced to their final state ready for absorption. The proteins have been broken down into peptones by the gastric and pancreatic enzymes, and into the *polypeptides* and *amino-acids* by the action of the succus entericus. Fats have been reduced to *fatty acids* and *glycerin*. Carbohydrates have been finally broken down into the *monosaccharides*, the main one, *glucose*, being very easily absorbed.

The digested food reaches the end of the small intestine in about four hours.

Absorption. The absorption of digested food takes place entirely in the small intestine through two channels, the capillary blood vessels and the lymphatics of the villi on the inner surface of the small intestine.

A *villus* contains a lacteal, blood vessels, epithelium, and muscular tissue, which are connected together by *lymphoid tissue* (*see* Fig. 14/11). The *central lacteal* ends in a blind extremity, plain muscle tissue lies along it, and it is surrounded by capillary blood vessels. The whole is then enclosed in a basement membrane and covered by epithelium. As the villi project from the intestinal wall, they are in contact with the liquid food or chyme and the fats are absorbed into the lacteals. The

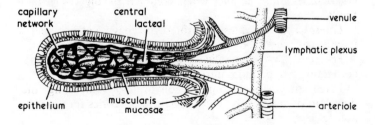

FIG. 14/11.—THE STRUCTURE OF A VILLUS

Summary of Digestive Processes

Organ	Digestive Fluid	Reaction	Enzymes	Chemical Action of Enzymes
Mouth	Saliva	Alkaline	Ptyalin (Salivary Amylase)	Converts cooked starches into a soluble sugar—Maltose
Stomach	Gastric Juice	Acid	(1) Rennin	Converts caseinogen into Casein
			(2) Pepsin	Converts proteins into Peptones
			(3) Gastric Lipase	Begins the hydrolysis of fats
Duodenum. . .	Bile	Alkaline	—	Aids action of pancreatic enzymes. Emulsifies fats
,,	Pancreatic Fluid	Alkaline	(1) Trypsin	Reduces proteins and peptones into polypeptides and Amino-acids
			(2) Amylase	Converts all sugars and starches into Maltose
			(3) Lipase	Reduces fats to Glycerin and Fatty-acids
Small Intestine	Succus Entericus	Alkaline	(1) Enterokinase	Sets free the trypsin in pancreatic fluid
			(2) Erepsin	Reduces all protein substances to Amino-acids
			(3) {Sucrase Maltase Lactase	Reduce all carbohydrate substances into the monosaccharides, Glucose, Galactose, and Laevulose

Summary of Absorption

Source of Food	Final Digested Product	Organ of Absorption
Proteins . . .	Amino-acids	Epithelium of villi into blood vessels and blood stream
Fats.	Glycerin and fatty acids	Epithelium of villi into lacteals and lymph stream
Carbohydrates .	Monosaccharides: Glucose Laevulose Galactose	Epithelium of villi and walls of blood vessels into blood stream

absorbed fats then pass by numerous lymphatic vessels to the *receptaculum chyli*, and thence by the *thoracic duct* to the blood stream (*see* page 199).

All other digested foods pass directly into the capillary blood vessels of the villi, and are carried by the portal vein to the liver, where certain changes take place (*see* pages 245–6).

THE LARGE INTESTINE AND DEFAECATION

The large intestine or colon, which is about 1·5 m (5 feet) long, is continuous with the small intestine at the *ileocolic* or *ileocaecal valve* through which the food residue passes. A *gastrocolic reflex* occurring when food enters the stomach excites peristalsis in the large intestine. It is this reflex which brings about defaecation (*see* page 235).

The colon begins as a dilated pouch, the *caecum*, to which the *vermiform appendix* is attached. The *appendix* is composed of the same four coats as the intestine but the sub-mucous coat contains a considerable amount of lymphoid tissue, which is thought to have a function similar to that of the tonsils. It may lie below or behind the caecum; in the latter case it is described as retrocaecal. The appendix becomes inflamed in *appendicitis* which generally necessitates the operation of appendicectomy.

The *caecum* lies in the right iliac region resting on the ilio-psoas muscle, and from here the colon ascends through the right lumbar region as the *ascending colon*. It turns beneath the liver as the *hepatic flexure*, passes across the margins of the epigastric and umbilical regions as the *transverse colon*, turns beneath the spleen as the *splenic flexure*, and passes down through the left lumbar region as the *descending colon*. In the left iliac region a bend called the *sigmoid flexure* or *pelvic colon* is formed, and it then enters the true pelvis and becomes the *rectum* (*see* Fig. 14/9, page 227).

The *rectum* is the lowest 13 cm (5 inches) of the large intestine, it begins at the pelvic colon and ends in the *anal canal* which is about 4 cm (1½ inches) long. This ends in the *anus*, which is guarded by internal and external sphincter muscles.

Structure. The *colon* consists of the same four coats as the small intestine, the longitudinal fibres of the muscular coat are arranged in three bands which give the colon a puckered and sacculated appearance. The inner mucous coat is smoother than that of the small intestine, it has no villi. It contains glands similar to the tubular intestinal glands and is lined by columnar

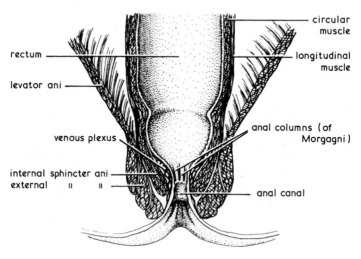

Fig. 14/12.—Rectum and Anal Canal, Coronal Section

epithelium which contains many secreting or goblet cells.

The structure of the *rectum* is similar to that of the colon, but the muscular coat is thicker and the mucous membrane is arranged in longitudinal folds called the *columns of Morgagni*. These are continued into the anal canal where the circular muscle fibres are thickened to form the *internal anal sphincter*. The cells lining the anal canal change in character, stratified epithelium replaces columnar cells.

The *external sphincter* keeps the anal canal and orifice closed.

The Functions of the Large Intestine. The large intestine does not take part in the digestion or absorption of food. When the contents of the small intestine reach the caecum all the nourishment has been absorbed, and the contents are liquid. In passing along the colon the contents become more solid as water is absorbed, and when the rectum is reached the faeces are of a soft-solid consistence. Peristalsis is very slow in the colon. It takes about sixteen to twenty-four hours for the contents to reach the sigmoid flexure.

The functions of the colon may be summarized as follows:

The *absorption* of water, salt and glucose,

The *secretion of mucin* by the glands in the inner coat,

The *preparation of cellulose* which is a carbohydrate present in plants, fruits and green vegetables, and any undigested protein is prepared by bacterial action for excretion.

Defaecation. The rectum is normally empty until just before defaecation. In a person of regular habits the call to defaecate occurs at about the same time each day. This is brought about by the gastro-colic reflex (*see* page 233), which usually functions after breakfast. This meal having reached the stomach and digestion begun peristalsis is stimulated in the intestine, spreads to the colon, and the residue from the previous day's food, which during the night has reached the caecum, begins to move. The contents of the pelvic colon enter the rectum; simultaneously strong peristalsis occurs in the colon and perineal sensation is experienced. Intra-abdominal pressure is increased by closure of the glottis and contraction of the diaphragm and the abdominal muscles; the anal sphincters relax, and the act is complete.

The *act of defaecation* is a matter of habit. Children are taught to defaecate after breakfast, before the interests of the day may cause the act to be inhibited, thus leading to habits of constipation. Some people defaecate before breakfast, others after; some who leave home early defaecate after arrival at work; others in the evening when there is leisure to attend to their needs. Some defaecate once a day or more often, others every other day, or at longer intervals. It should never be thought that there is a 'correct' time or frequency; people vary.

Composition of faeces. Faeces contain a considerable quantity of bacteria, most of them dead, shed epithelium from the intestine, a small quantity of nitrogenous matter, mainly mucin; also salts, principally calcium phosphate, and a little iron, cellulose and any other undigested food residue and water.

THE PERITONEUM

The peritoneum is a double *serous membrane*, the largest in the body, consisting of two main parts, the *parietal peritoneum* which lines the walls of the abdominal cavity, and the *visceral peritoneum* which is reflected over the organs contained in that cavity. The potential space between these two layers is called the *peritoneal space* or sac. In the male it is a closed sac; in the female the uterine (Fallopian) tubes open into the peritoneal cavity. Many sacs and folds extend from the peritoneum; one large fold is the *greater omentum* richly supplied with fat hanging down from the greater curvature of the stomach.

The *lesser omentum* descends from the *porta hepatis* after enclosing the liver, to the lesser curvature of the stomach where it splits to encircle this organ. The transverse colon is also enclosed within this peritoneum which then passes upwards and backwards as the *meso-colon* to the posterior abdominal wall. Part of this peritoneum forms the mesentery of the small intestine.

The greater and lesser omenta, the mesentery of the small intestine and the meso-colon all carry the vascular and lymphatic drainage from the organs they encircle.

The Functions of the Peritoneum. It covers most of the abdominal and pelvic organs, forming a smooth lining which enables these organs to move upon each other without friction.

It attaches the organs together and keeps them in position

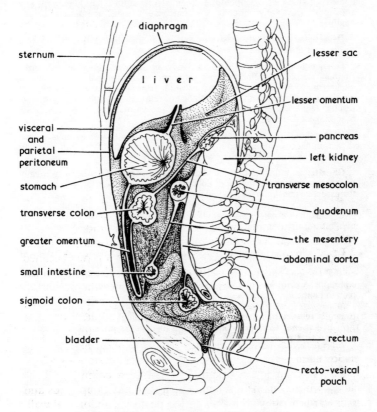

FIG. 14/13.—DIAGRAM OF THE PERITONEUM

The omenta are folds of peritoneum separating some of the abdominal organs. The mesenteries join parts of the intestine together and sling them up to the posterior abdominal wall.

and maintains the organs in relation to the posterior abdominal wall.

The numerous lymphatic nodes and vessels with which the peritoneum is supplied help in protecting it from infection.

Clinical Notes

The Alimentary Tract. A number of conditions may arise in various parts of the alimentary canal—a few are mentioned below. Certain symptoms are common to most of these disorders or diseases, e.g. nausea, coated tongue, loss of appetite, abdominal discomfort, pain, vomiting and constipation, varying in character and severity with the nature of the condition.

Dysphagia, a difficulty in swallowing most often associated with disease or disorder of the throat or oesophagus.

Oesophagus. *Stenosis or stricture* may be due to injury as in burns, tumour, simple or malignant, pressure from a mediastinal tumour, or an aneurysm within the thorax; or without from an enlarged thyroid. In *cardiospasm* the cardiac sphincter does not relax as it should and dysphagia results. *Oesophagitis*, or inflammation of the oesophagus, may complicate oesophageal hiatus hernia (*see* p. 148).

Dyspepsia. A form of indigestion often persistent, which may be due to a number of causes, including errors of diet, irregular mealtimes and sometimes associated with anxiety and apprehension.

Gastritis is inflammation of the stomach. Acute gastritis is generally due to an irritant, as in food poisoning, infection such as influenza and the excessive consumption of alcohol.

The *symptoms* are those common to disorders of the alimentary canal (*see* above). *Chronic gastritis* is a term employed to describe chronic indigestion, or dyspepsia (*see* above) occurring in middle-aged and elderly people.

Peptic Ulcer occurs on parts of the stomach and duodenum exposed to the action of gastric juice. There are many causes, including irregular meals, tension, anxiety and emotional stresses. The *symptoms* are those common to disorders of the alimentary tract (*see* above), except that the pain in peptic ulcer varies with the site of the ulcer.

In a gastric ulcer pain comes on about 20 minutes after food and is generally relieved by taking an antacid or by vomiting.

In a duodenal ulcer pain begins about two hours after food. A patient may say that it wakens him after sleeping for a couple of hours and that it can be relieved by food, a few biscuits, a milky drink or by taking an antacid mixture.

Treatment. Peptic ulcer may respond to conservative medical treatment, by diet, rest and antacids, with, if possible, freedom from anxiety and responsibility. When the condition does not respond within a reasonable period, surgical measures are considered.

Pyloric Stenosis. A contracted pyloric sphincter may be congenital, when it may be relaxed by drugs or require surgical intervention. Pyloric stenosis in adults may complicate a duodenal ulcer.

The *vomiting in pyloric stenosis* is projectile in character, through nose and mouth, even when infants are being correctly fed, so that fluid and electrolytes are lost and there is dehydration and wasting.

Enteritis or inflammation of the large or small intestine, is often associated with acute gastritis or *gastro-enteritis*, and in many instances is due to infection, as in bacterial food poisoning.

In infants there is excessive diarrhoea and vomiting with dehydration, a low blood pressure and severe prostration. The *treatment* is appropriate antibiotics, intravenous injection of the correct fluids containing electrolytes until vomiting ceases, when water, whey, and diluted milk may be given. *Isolation is essential.*

In adults gastro-enteritis is frequently due to food poisoning, particularly to Salmonella infections.

Colic is an intermittent acute pain due to powerful contractions of the muscular walls of hollow viscera. **Gastro-intestinal colic** can be very severe. The patient, extremely restless, is doubled up and rolling about in pain. He suffers from shock and prostration, his pupils are dilated by the pain, he is anxious and frightened.

Treatment. Adequate sedation to relieve pain; applications of heat to the abdomen may help.

Colitis or inflammation of the colon. *Ulcerative colitis* is characterized by marked ulceration and dilatation of the colon, with the passage of watery offensive stools containing blood and mucus. A patient can be extremely ill and a subject of invalidism for years, not responding to any medical treatment, so that surgery is considered and *colectomy* performed.

Ileostomy, which is permanent, is then carried out—an opening is made into the terminal end of the small intestine—ileum—which acts as an artificial anus. This patient needs careful nursing, encouragement and education in the management of his ileostomy. He is faced with making physical and mental adjustments in this new situation, but it can be done and the advantage of good health lies before him. A patient should know that any *excessive liquid efflux* must be referred to his doctor, as this loss of fluid and electrolytes should be dealt with immediately.

Colostomy is an opening made through the abdominal wall into the iliac (ascending) colon, through which faeces are evacuated. *Colostomy may be temporary* when it will be closed later, or it may be *permanent*, acting as an artificial anus after excision of the rectum. The important points in care are to aim at an evacuation daily; to regulate bowel action by dietary changes, not by aperients. If diarrhoea is disturbing, it may be checked by arrowroot or a kaolin and chlorodyne mixture might be needed.

Abdominal catastrophes are common surgical emergencies and are due, for instance, to an *inflammation* such as appendicitis (one of the commonest abdominal emergencies) or cholecystitis, which may spread to the peritoneum causing *peritonitis* (*see* p. 240). *Intestinal obstruction* is due to narrowing of the bowel lumen. This may be caused by blockage of the bowel lumen from within, constriction of the wall as in a malignant growth, or by outside pressure and constriction as in a strangulated hernia. *Other acute abdominal conditions* are due to the *perforation* of an organ.

Peritonitis is inflammation of the peritoneum which may be generalized or local. *Acute general peritonitis* may follow perforation of one of the hollow viscera, or of the appendix or gall-bladder. There is acute abdominal pain, a rigid distended abdomen and shock, with a rapid thready pulse and shallow breathing. Vomiting and hiccup may be distressing. Surgical measures are needed, antibiotics and supportive therapy—blood transfusions, oxygen administration as required.

Ascites is a collection of fluid in the peritoneal cavity occurring as the result of heart or liver failure. It is seen in the later stages of congestive heart failure and when the function of the liver is seriously disordered in carcinoma.

Malignant disease may affect any part of the alimentary tract, but more particularly the oesophagus, stomach, colon and rectum.

Vomiting may occur as the result of disorder of any part of the alimentary tract, in poisoning, due to motion, as in travel sickness, as the result of pain, fear and for many other reasons. There is nausea, increased salivation, some quick irregular breathing, with retching. This is due to contraction of the respiratory muscles, contraction of the diaphragm, and closure of the glottis whilst the vomited matter is ejected. These movements are controlled by a vomiting centre in the medulla.

A patient who is vomiting often feels faint and cold, beads of perspiration collect on his face; he needs support and comforting.

Vomiting in pyloric stenosis and in some cerebral lesions is projectile in character.

Constipation or diminished bowel output has many causes. The diet may be deficient in fat, water, fruit or vegetables; faulty training in bowel habit in children or neglect of nature's call to defaecation later; the habitual use of aperients, suppositories or enemas results in the bowel not acting without this aid. Constipation may also be associated with indigestion, or spasticity of the colon, the existence of a tumour or other cause of obstruction.

(*See also* Dental hygiene, page 216. Disorders of tonsils, page 217. Disorders of salivary glands, page 220. Examination of stomach, page 225.)

Chapter 15

THE LIVER, GALL-BLADDER
AND PANCREAS

THE LIVER

The liver is the largest gland in the body, situated in the uppermost part of the abdominal cavity on the right side beneath the diaphragm. It is largely protected by the ribs.

The liver is divided into two main lobes, right and left. The upper surface is convex and lies beneath the diaphragm; the under surface is irregular and presents the *transverse fissure*, the surface being broken by the passage of the vessels which enter

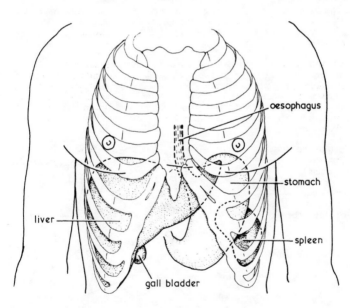

Fig. 15/1.—The Position of the Liver shown rising as high as the Fifth Rib, and extending as low as the Costal Margin on the Right Side

241

and leave the liver. The *longitudinal fissure* separates the caudate and left lobes on the under surface and the *falciform ligament* occupies a similar position on the upper surface of the liver. The liver is further subdivided into four lobes (right, left, caudate and quadrate) (*see* Fig. 15/3). These are made up of lobules. The lobules, which are polyhedral in shape, are composed of cubical liver cells and the ramifications of the vessels of the liver, all connected by liver tissue. The liver has a double blood supply by means of the hepatic artery and the portal vein.

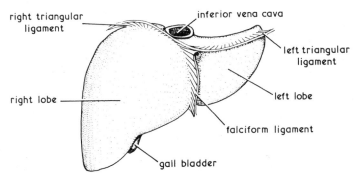

FIG. 15/2.—THE ANTERIOR SURFACE OF THE LIVER

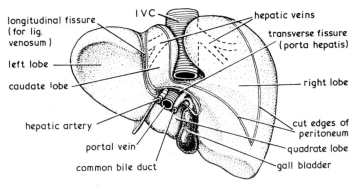

FIG. 15/3.—THE BACK AND UNDER SURFACES OF THE LIVER

The Vessels of the Liver are:

The *hepatic artery*, which arises from the aorta and supplies one-fifth of the blood to the liver; this blood has an oxygen saturation of 95 to 100 per cent.

The *portal vein*, which is formed from the splenic vein and the superior mesenteric vein, supplies four-fifths of the blood to the liver; this blood has an oxygen saturation of only 70 per cent because some O_2 has been taken up in the spleen and intestine. This portal (venous) blood brings to the liver the nutrients absorbed by the mucosa of the small intestine.

The *hepatic vein* returns the blood from the liver to the inferior vena cava; there are no valves in the hepatic veins.

Bile ducts are formed by the union of the bile capillaries which collect the bile from the liver cells.

There are thus four main vessels traversing the substance of the liver, two entering, the hepatic artery and portal vein, and two leaving, the hepatic vein and the bile duct.

Minute Structure. The liver cells are nucleated polyhedral cells. The protoplasm of the cells contains large numbers of enzymes. Masses of these cells form the hepatic lobules which are roughly hexagonal in shape (*see* Fig. 15/5), about one millimetre in diameter and separated from one another by a connective tissue in which run the ramifications of the vessels

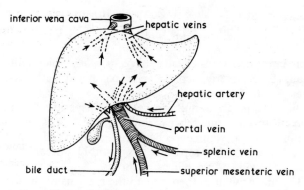

FIG. 15/4.—THE VESSELS ENTERING AND LEAVING THE LIVER

traversing the liver. Branches of the portal vein, the hepatic artery and the bile ducts are enclosed together in a connective tissue covering, called Glisson's capsule, which forms portal canals. The blood from the portal vein comes into close proximity with the liver cells, each lobule is penetrated by a network of *blood sinusoids* or hepatic capillaries (*see* Fig. 15/5). Small vessels passing between the liver lobules are called *interlobular veins*. From these, capillaries branch into the substance of the lobules, and unite to form a small vein in the centres of the lobules, *intralobular veins*. These vessels pour their con-

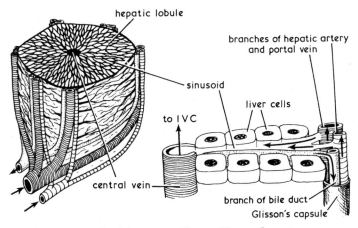

FIG. 15/5.—MICROSCOPIC VIEW OF HEPATIC LOBULE

Note the hexagonal shape. In the tissue surrounding the lobule lie branches of the portal vein, hepatic artery, bile ducts and the lymphatics.

tents into other veins called *sublobular veins*, which unite and finally form several hepatic veins passing directly into the inferior vena cava.

Bile is formed in minute spaces in the hepatic cells and discharged through fine bile capillaries or *bile canaliculi*, minute channels which commence between the liver cells, lying between two cells, but separated from the blood capillaries so that the blood and the bile never mix.

The bile capillaries then pass to the margins of the lobules, and pour their contents into the interlobular bile ducts which unite to form the *hepatic ducts*. The largest bile ducts are lined by columnar epithelium and have an outer coat of fibrous and muscular tissue; by means of the contraction of the muscular covering of these ducts bile is carried away from the liver.

The Functions of the Liver are concerned with the metabolism of the body, particularly by its action on food and blood.

The liver is the largest chemical factory in the body in that it carries out most of the 'intermediate metabolism', that is, it modifies the nutrients absorbed from the gut and stored elsewhere in the body to make them suitable for use in the tissues.

The liver also modifies waste products and toxic substances to make them suitable for excretion in the bile or the urine.

The glycogenic function. Stimulated by the action of an enzyme the liver cells produce glycogen (an animal starch) by concentration of the glucose derived from the carbohydrate food. This substance is stored temporarily by the liver cells, and converted back into glucose by enzyme action when needed by the body tissue. By means of this function the liver aids in maintaining the normal blood sugar level of 80 to 100 mg glucose per 100 ml of blood, but this is controlled by the internal secretion of the pancreas, insulin (*see* page 282). The liver can also convert amino-acids into glucose.

The secretion of bile. Some of the constituents of bile, e.g. bile salts, are made in the liver; other constituents, e.g. bile pigments, are formed in the reticulo-endothelial system and passed into the bile by the liver.

Formation of urea. The liver receives the amino-acids which have been absorbed by the blood. In the liver cells deamination takes place, which means that the nitrogen is separated from the amino-acid part, and the ammonia is converted into urea. Urea is eventually removed from the blood by the kidneys and excreted in the urine.

Action on fats. The liver prepares the fats for their final breaking down into the end products of carbonic acid and water. The bile salts produced by the liver are essential for the digestion and absorption of fats. Any decrease in bile salts reduces the absorption of fat which may then pass unaltered into the faeces as occurs in some digestive disorders of young children, in coeliac disease, tropical sprue and certain disorders of the pancreas.

The liver is also concerned with the **normal content of blood:**
(*a*) It forms red cells in fetal life.
(*b*) It plays a part in the destruction of red blood cells.
(*c*) Stores the haematin needed for the maturation of new red cells.
(*d*) Manufactures most of the plasma proteins.
(*e*) Removes bilirubin from the blood.
(*f*) Is concerned with the production of prothrombin and fibrinogen essential for the clotting of blood (*see* page 170).

Storage and distribution of many substances including glycogen, fat, vitamins and iron. The fat soluble vitamins, A and D, are stored in the liver which is why liver oils are such a good source of these substances.

Maintenance of body temperature. The liver helps to maintain the temperature of the body because of its size and the number of its *metabolic activities* which cause the blood passing through the organ to be raised in temperature.

The protective action of the liver is also described as detoxication. Some of the barbiturate drugs, and alcohol, can be completely destroyed by the liver; but poisoning by large doses of hypnotic drugs may damage the liver cells. Similarly some of the chemicals used in industry, such as carbon tetrachloride, cause damage, and careful watch is kept on the effect of recent chemical preparations and drugs on the market regarding their effects on the liver.

THE GALL-BLADDER

The gall-bladder is a pear-shaped musculo-membranous bag, lying in a fossa on the under surface of the liver and reaching to the front margin of that organ. It measures 8–10 cm (3–4 inches) in length and holds about 60 ml.

It is divided into a fundus, body, and neck, and consists of three coats:

An Outer serous peritoneal coat,
A Middle unstriped muscular tissue, and
An Inner mucous membrane, which is continuous with that
lining the bile ducts. The mucous membrane is comprised of

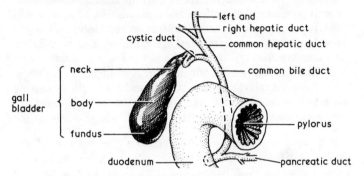

left and
right hepatic duct
cystic duct
common hepatic duct
neck
common bile duct
gall bladder
body
pylorus
fundus
duodenum
pancreatic duct

Fig. 15/6.—Diagram of the Gall-Bladder and Bile Ducts

columnar epithelial cells which secrete mucin and rapidly
absorbs water and electrolytes, but not bile salts or pigments,
thus concentrating the bile.

The cystic duct is about 4 cm (1½ inches) in length. It
passes from the neck of the gall-bladder and joins the hepatic
duct, thereby forming the common bile duct which conveys
the bile to the duodenum.

Function. The gall-bladder acts as a *reservoir* for the bile.
It also performs the important function of *concentrating the
bile* stored in it.

Within half an hour of taking food, as the sphincter of Oddi
relaxes to permit bile to enter the duodenum, the gall-bladder
contracts; thus the flow of bile is not continuous but corresponds with the intervals during digestion when food enters
the duodenum.

Composition and Function of Bile. Bile is an alkaline fluid
secreted by the liver cells. The amount secreted daily in man is
from 500 to 1,000 ml, the secretion is continuous, but the rate
of production is accelerated during digestion, especially during
the digestion of fats (*see* page 230). Bile contains about 86 per

cent water, bile salts, bile pigments, cholesterol, mucin, and other substances. *Choleretics* increase the secretion of bile. *Cholagogues* cause emptying of the gall-bladder.

Bile pigments. These are formed in the reticulo-endothelial system (particularly the spleen and bone marrow) by the breakdown of haemoglobin from destroyed red cells, which are passed to the liver where they are excreted into bile. These pigments are conveyed by the bile to the small intestine; some of it becomes *stercobilin*, which colours the faeces, and some is re-absorbed into the blood stream and forms the colouring matter of urine, *urobilin*. Bile pigments are merely excretory products. They have no digestive action.

Bile salts are digestive and facilitate the activity of the fat-splitting ferment lipase. The bile salts also aid in the absorption of digested fat (glycerin and fatty acids) by lowering the surface tension and increasing the permeability of the endothelium covering the intestinal villi.

THE PANCREAS

The pancreas is a compound racemose gland, very similar in structure to the salivary glands. It is about 23 cm (7 inches) long, extending from the duodenum to the spleen, and is described as consisting of the following three parts.

The head of the pancreas, the broadest part, lies to the right of the abdominal cavity and in the curve of the duodenum, which practically encircles it.

The body of the pancreas is the main part of the organ; it lies behind the stomach and in front of the first lumbar vertebra.

The tail of the pancreas is a narrow part to the left, which actually touches the spleen.

The substance of the pancreas is composed of lobules of secretory cells arranged round tiny ducts, which begin by the junction of the small ducts of lobules situated in the tail of the pancreas, passing through the body from left to right, receiving ducts from other lobules and uniting to form the main duct, the *duct of Wirsung* (*see* page 250).

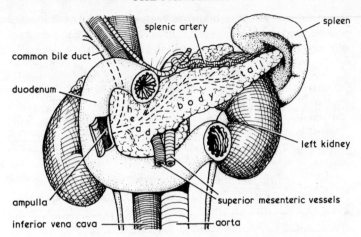

FIG. 15/7.—THE PANCREAS AND ITS RELATIONS (STOMACH REMOVED)
The head of the pancreas is encircled by the duodenum, and the tail touches the spleen.

Functions. The pancreas, which may be described as a dual organ, has two functions. The *exocrine function* is carried out by the secretory cells of its lobules, which form pancreatic juice containing enzymes and electrolytes. This digestive fluid passes through tiny excretory ducts and is finally col-

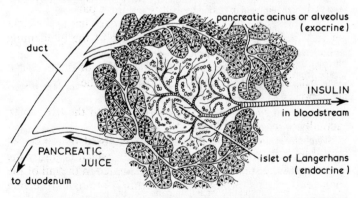

FIG. 15/8.—FROM A MICROSCOPIC SECTION OF PANCREAS SHOWING ISLET OF
LANGERHANS SURROUNDED BY ALVEOLI

lected by two ducts, a main one called the *duct of Wirsung*, and an accessory duct, of *Santorini*, which open into the duodenum. The main duct joins the bile duct at the *ampulla of Vater* (*see* Fig. 15/7). The enzyme content of the pancreatic juice is mentioned on page 230, where its digestive function is given. The pancreas is supplied by the vagus nerves, and within a few minutes of taking food, the flow of pancreatic juice is increased. Later, when the gastric contents pass into the duodenum two hormones, *secretin* and *pancreozymin*, are formed in its mucosa which further stimulate the flow of pancreatic juice.

The *endocrine function*. Scattered between the alveoli of the pancreas (*see* Fig. 15/8) are small groups of epithelial cells, quite separate and distinct. These are the *Islets of Langerhans* which collectively form an endocrine organ described in Chapter 18, page 282. Their nerve supply is from the vagus and their blood supply from large capillary loops.

Clinical Notes

The *liver* may be ruptured by injury or penetrated by a broken rib and either may cause serious bleeding; it may be the site of abscess, hydatid cyst, toxic degeneration and *cirrhosis*. Its function may be impaired in the late stages of congestive heart disease and also in cancer, when *jaundice*, vomiting, ascites (fluid in the peritoneal cavity) and portal (liver) congestion, bring their train of discomforts.

The *gall-bladder* is liable to infection, *cholecystitis*, which may be conveyed to it either from the intestine or liver or through the blood stream. *Gall stones* may be formed in the gall-bladder and when these obstruct the hepatic or the common bile duct (*see* Fig. 15/6) the bile cannot escape from the liver and *obstructive jaundice* occurs. Stones in the gall-bladder and cystic duct do not give rise to jaundice. If a gall stone passes down the bile duct it may cause severe pain, *biliary colic*.

Jaundice may be due to one of three main causes.
First, the production of too much bile pigment, as in haemolytic anaemia.
Second, failure of the liver cells to modify or excrete the bile, as in infective hepatitis.
Third, obstruction of the common bile duct as by a gall stone, as mentioned above, or a carcinoma of the head of the pancreas.

The *pancreas*. Pancreatitis or inflammation of the pancreas is associated with disorders of the biliary tract; the condition may be acute or chronic; it is characterized by pain, distaste for food, nausea and vomiting and, when acute, by prostration. Some degree of pancreatic insufficiency often follows and diabetes may supervene.

Cancer of the pancreas may cause jaundice, failure of digestion with steatorrhoea, or diabetes. The condition is difficult to treat surgically, but complete removal of the pancreas (pancreatectomy) may be possible; if so, it is followed by mild diabetes (which is treated with insulin); or mild malabsorptive syndrome may develop, which is characterized by loss of weight, increased excretion of fat in the stools, some anaemia, and treated by taking pancreatic digestive enzymes by mouth.

(For disorder of the function of the internal secretion of the pancreas, *see* page 282.)

Chapter 16

THE RESPIRATORY SYSTEM AND RESPIRATION

It is by means of **breathing** that every cell in the body receives its supply of oxygen and at the same time gets rid of the products of oxidation. Oxygen combining with the carbon and hydrogen of the tissues enables the metabolic processes of each individual cell to proceed, with the result that work is effected and waste products in the form of carbon dioxide (CO_2) and water (H_2O) are eliminated.

Respiration is a two-fold process whereby the *interchange of gases* takes place in the tissues, 'internal respiration'—and in the lungs, 'external respiration' (for description of the physiology of respiration, *see* page 261).

Air is drawn into the lungs during inspiration and expelled from the lungs during expiration. The air enters through the respiratory passages which are enumerated and briefly described below.

The bony framework of the nose is described on page 72, and the olfactory region on page 369. The respiratory portion forms the upper part of the respiratory passages.

THE RESPIRATORY PASSAGES

The **anterior nares** are the openings into the nostrils. They open into the portion known as the *vestibule of the nose* which is lined with stratified epithelium continuous with that of the skin. The lining of the anterior nares contains a number of sebaceous glands and is covered by coarse hairs. They open into the nasal cavities.

The **nasal cavities** are lined with mucous membrane which is highly vascular. It is continuous with the lining of the pharynx and with the mucous membrane of the sinuses which open into the nasal cavity. The respiratory region is lined with columnar and ciliated epithelium containing goblet or mucous cells. The secretion of these cells renders the surface moist and sticky. The membrane is thickest over the nasal septum and the conchae, which are mentioned below. The presence of three turbinate bones (conchae) which are covered with respiratory epithelium and project from the lateral wall of the nose into the cavity, greatly increase its surface area.

As air passes through the nose it is *filtered* by the hairs contained in the vestibule, *warmed* by contact with the fairly

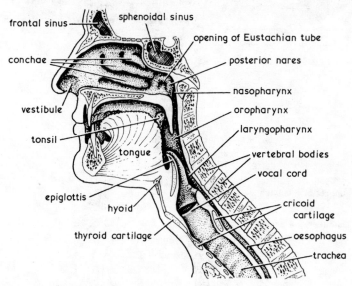

FIG. 16/1.—SECTION OF THE FACE AND NECK, SHOWING THE UPPER RESPIRATORY PASSAGES

extensive mucous surface, and *moistened* by evaporation of moisture from this surface. The nose receives the openings from the para-nasal air sinuses which therefore drain into the nasal cavities, and the opening of the naso-lacrimal duct which passes the tears from the eyes into the lower part of the nasal cavities.

The **pharynx** is a muscular tube which extends from the base of the skull to its junction with the oesophagus at the level of the cricoid cartilage. It therefore lies behind the nose (the *naso-*

pharynx), behind the mouth (the *oro-pharynx*), and behind the larynx (the *laryngeal pharynx*).

The *posterior nares* are the openings from the nasal cavities to the naso-pharynx.

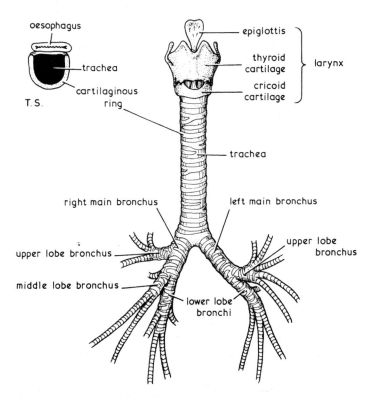

FIG. 16/2.—THE LARYNX, TRACHEA, AND BRONCHI, WITH THEIR MAJOR BRANCHES

The **larynx** lies in front of the lowest part of the pharynx which separates it from the vertebral column, extending from the pharynx to the level of the 6th cervical vertebra. It is the upper prominent part of the windpipe and opens into the trachea below.

The larynx is composed of pieces of cartilage connected together by ligaments and membrane. The largest of these is the *thyroid cartilage*, the front of which forms the subcutaneous prominence known as Adam's apple in the front of the neck. It consists of two plates or laminae joined in the middle line. The superior border is marked by a V-shaped notch. The *cricoid cartilage* lies below the thyroid and is shaped like a signet ring with the signet part of the circle at the back. (This is the only cartilage which is a complete ring.) Other cartilages are the two *arytenoid cartilages* perched on the back of the cricoid and the paired *cuneiform* and *corniculate* cartilages, which are very small.

Attached to the top of the thyroid cartilage is the *epiglottis*, which forms a cartilaginous flap and helps to close off the larynx during swallowing. The larynx is lined with the same type of mucous membrane as the trachea except that the vocal cords and part of the epiglottis are covered by stratified epithelial cells.

The *vocal cords* lie inside the larynx, passing from the thyroid cartilage in front to the arytenoids behind. By movements of the arytenoid cartilages brought about by various laryngeal muscles the vocal cords are *approximated* or *relaxed*. Thus the size of the opening between the cords, the *rima glottidis*, is altered during breathing and speech.

It is by the vibration of the cords due to the air passing through the glottis that the voice is produced. Various muscles attached to the larynx control the production of voice, and also close the upper opening of the larynx during swallowing.

The Trachea. The trachea or windpipe is about 10 cm (4 inches) long. It extends from the larynx to about the level of the 5th thoracic vertebra, where it divides into two bronchi. It is composed of sixteen to twenty incomplete rings of cartilage connected by fibrous tissue which completes the circumference at the back of the trachea; in this situation it contains some muscular tissue. The trachea is lined by mucous membrane composed of ciliated epithelium and goblet cells. The direction in which the cilia move is upwards towards the larynx, and by this movement inhaled particles of dust, pollen, etc. are expelled. The cartilages which serve to keep the trachea open are incomplete behind where the trachea is in close contact with the oesophagus, which separates it from the vertebral column.

The *cervical trachea*, which passes through the neck, is crossed by the isthmus of the thyroid gland, the lobes of the gland embracing the sides of the trachea. The *thoracic trachea* passes through the superior mediastinum lying behind the sternum, in contact with the brachiocephalic artery and the arch of the aorta. The oesophagus lies behind the trachea.

The Bronchi, which are formed by the bifurcation of the trachea at about the level of the 5th thoracic vertebra, are similar in structure to the trachea, and are lined by the same types of cells. The bronchi pass downwards and outwards towards the roots of the lungs. The *right bronchus* is shorter and wider than the left; it gives off one branch at a level higher than that of the pulmonary artery called the *upper* lobe bronchus; the other branch arising after the main branch has passed below the artery is the *lower lobe* bronchus. The *middle lobe* bronchus arises from the lower lobe bronchus (*see* Fig. 16/2).

The *left bronchus* is longer and slimmer than the right; it passes below the pulmonary artery before dividing into branches to the upper and lower lobes.

THE THORACIC CAVITY

The thoracic cage which is a bony and cartilaginous cavity, has been described on page 74. *The boundaries which convert the thorax into a cavity* are:

The sternum and costal cartilages in front,
The twelve thoracic vertebrae with their intervertebral discs of cartilage behind,
The ribs and intercostal muscles at the sides,
The diaphragm below, and
The root of the neck above.
Contents. The sides of the thoracic cavity are completely filled by the lungs with their pleural covering; these lie each side of, and form the lateral boundaries of, the mediastinum.

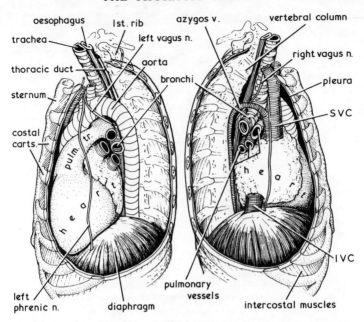

FIG. 16/3.—THE MEDIASTINUM SEEN FROM EACH SIDE, THE LUNGS HAVING BEEN
REMOVED

The mediastinum is the space in the thoracic cavity between
the two lungs. It contains the heart and great blood vessels,
the oesophagus, thoracic duct, descending aorta, and superior
vena cava, the vagi and phrenic nerves and numerous lymphatic
glands.

THE LUNGS

The lungs, two in number, are the principal organs of respiration. They fill the chest cavity, lying one on each side
separated in the middle by the heart and its great blood vessels,
and by the other structures lying in the mediastinum (*see*
above). The lungs are cone-shaped organs, with the *apex*
above, rising a little higher than the clavicle into the root of
the neck. The *base of the lungs* lies resting on the floor of the
thoracic cavity, on the diaphragm. The lungs present an *outer*

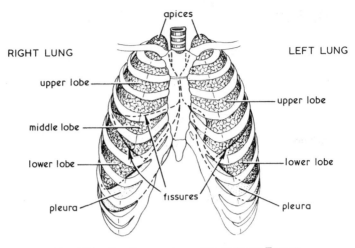

FIG. 16/4.—THE POSITION OF THE LUNGS IN THE THORAX

surface in contact with the ribs, an *inner surface* where the root of the lung lies, a *posterior border* in contact with the vertebral column, and an *anterior border* which overlaps the anterior aspect of the heart.

The Lobes of the Lungs. The lungs are divided into lobes by fissures. The right lung has three and the left lung two lobes. Each of these lobes is composed of a number of lobules. A small bronchial tube enters each lobule and as it divides and subdivides its walls become thinner and thinner and finally end in small dilated sacs, the air sacs of the lungs. Lung tissue is elastic, porous and spongy. It floats in water because of the air contained in it.

The Pulmonary Bronchi. The trachea divides into two main bronchi; these divide again before they enter the lungs (*see* page 254). As the *pulmonary bronchi* pass through the lungs they divide and subdivide a great number of times. The larger vessels retain a structure similar to that of the trachea, having a fibrous muscular wall containing cartilage and lined by ciliated epithelium. The cartilage gradually disappears from

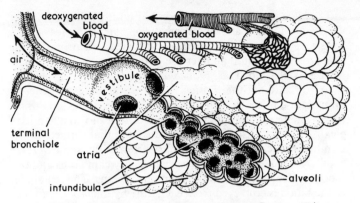

FIG. 16/5.—THE TERMINATION OF A BRONCHIOLE INTO GROUPS OF ALVEOLI
The capillary network covers all the alveoli but is here shown investing only three.

the smaller vessels leaving the fibro-muscular coat and the cilia-ted lining. The *terminal bronchioles* open into a slightly altered passage called the *vestibule*, and here the lining membrane begins to change its character; the ciliated epithelial lining gives place to one of flattened epithelial cells. From the vestibule (*see* Fig. 16/5) several *infundibula* open, and in the walls of these the air sacs lie. The *air sacs* or *alveoli* consist of one single layer of flat epithelial cells, and it is here that the blood comes into almost direct contact with the air—a plexus of capillary blood vessels surrounds the alveoli and the interchange of gases takes place.

The Blood Vessels of the Lungs. *The pulmonary artery* carries deoxygenated blood from the right ventricle of the heart to the lungs; its branches lie in contact with the bronchial tubes, dividing and subdividing until tiny arterioles are formed; these break up into a network of capillaries which lie in contact with the walls of the alveoli or air sacs.

These minute capillary vessels have a very small bore, so that the red blood cells are carried along practically in single file. They are moving slowly, and being separated from the air in the alveoli by only two exceedingly thin membranes the interchange of gases takes place by *diffusion*, which is the function of respiration (*see* page 261).

The pulmonary capillaries unite and unite again until larger vessels are formed and finally *two pulmonary veins* leave each lung carrying the oxygenated blood to the left atrium of the heart for distribution all over the body by means of the aorta.

Vessels described as *bronchial arteries* carry oxygenated blood direct from the thoracic aorta to the lungs in order to nourish and bring oxygen into the substance of the lung tissue itself. The terminal branches of these arteries form a capillary plexus, which is distinct and separate from that formed by the terminal branches of the pulmonary arteries, but some of these capillaries finally unite with the pulmonary capillaries and this blood is then carried into the pulmonary veins. The remainder of the blood is conveyed from each lung by the *bronchial veins*, and it eventually reaches the superior vena cava. The lungs therefore have a dual blood supply.

The Root of the Lung is formed by the following structures:

Pulmonary arteries, which carry deoxygenated blood into the lungs for oxygenation,

Pulmonary veins, which return the oxygenated blood from the lungs to the heart,

Bronchi, which branch into the bronchial tree and form the principal air passages,

Bronchial arteries, which arise from the thoracic aorta and convey arterial blood to the lung substance,

Bronchial veins, which return some of the blood from the lungs to the superior vena cava, and

Lymphatic vessels, which pass in and out of the lungs and are very numerous.

Nerves. The lungs are supplied by the vagus and sympathetic nerves.

Lymphatic glands. All the lymphatic vessels which pass through the lung structure are drained eventually into glands lying at the root of the lung.

The Pleura. Each lung is surrounded by a double serous membrane, the pleura. The *visceral pleura* closely invests the lung, passing into the fissures and so dividing the lobes from each other. This membrane is then reflected back at the root of the lung and forms the *parietal pleura*, which covers the interior of the chest wall. The pleura lining the ribs is the *costal pleura*, the portion covering the diaphragm the *diaphragmatic pleura*, and the portion which lies in the neck the *cervical pleura*. This is strengthened by a strong membrane called the *supra-pleural membrane* (Sibson's fascia), on which the subclavian artery lies.

Between the pleural layers there is a slight exudate which lubricates the surfaces, and prevents friction between the lungs and the chest wall during the respiratory movements. In health the two layers of pleura are in contact one with the other. The pleural space or cavity is only a potential space; but when, in abnormal states, air or fluid lies between the two layers of pleura separating them, the space then becomes distinct.

THE PHYSIOLOGY OF RESPIRATION

The **function of the lungs** is the interchange of the gases oxygen and carbon dioxide.

In **Pulmonary Respiration or External Respiration, oxygen** is taken in, through the nose and mouth, in **breathing;** it flows along the trachea and bronchial tubes to the alveoli, where it comes into intimate contact with the blood in the pulmonary capillaries. Only one layer of membrane, the *alveolar-capillary membrane*, separates the oxygen from the blood. *Oxygen* passes across this membrane and is taken up by the haemoglobin of the red blood cells and carried to the heart from whence it is pumped in the arteries to all parts of the body. Blood leaves the lungs at an oxygen pressure of 100 mm Hg and at this level the haemoglobin is 95 per cent saturated with oxygen.

In the lungs, **carbon dioxide,** a waste product of metabolism, passes across the alveolar-capillary membrane from the blood capillaries to the alveoli and, passing through the bronchial tubes and trachea, is breathed out through the nose and mouth.

Four processes are concerned in *pulmonary* or *external respiration:*

(1) *Pulmonary ventilation*, or the act of breathing, which replaces the air in the alveoli with outside air

(2) The *flow of blood through the lungs*

(3) The *distribution of air flow and blood flow* so that correct amounts of each reach all parts of the lungs

(4) *Diffusion of gases* passing across the alveolar-capillary membrane. CO_2 diffuses more readily than oxygen.

These processes are adjusted so that blood leaving the lungs has the correct amount of CO_2 and O_2. During exercise more blood comes to the lungs with too much CO_2 and too little O_2; the amount of CO_2 cannot be excreted and the concentration in the arterial blood increases. This stimulates the respiratory centre in the brain to increase the rate and depth of breathing. The *increased ventilation* thus brought about then excretes the CO_2 and takes up more O_2.

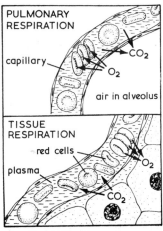

FIG. 16/6.—RESPIRATION

Tissue or Internal Respiration. The blood, having its haemoglobin saturated with oxygen (oxy-haemoglobin), circulates throughout the body and finally reaches the capillary bed where the blood is moving extremely slowly. The tissue cells take oxygen from the rich haemoglobin to enable oxidation to go on, and the blood receives in exchange the waste product of oxidation, carbon dioxide.

The following changes take place in the composition of air in the alveoli, brought about by external and internal or tissue respiration.

Inspired (atmospheric) *air:*

Nitrogen	79	per cent
Oxygen	20	,, ,,
Carbon dioxide	0·04	,, ,,

Air entering the alveoli is of the temperature and humidity of the atmosphere.

Expired air:

Nitrogen	79	per cent
Oxygen	16	,, ,,
Carbon dioxide	4·04	,, ,,

Expired air is saturated with water vapour and it is of the temperature of the body (20 per cent of the body heat is lost in warming the expired air).

Air Capacity of the Lungs. The *total air capacity* of the lungs is from 4,500 to 5,000 ml or 4½ to 5 litres of air. Only a small proportion of this air, about $\frac{1}{10}$ (500 ml) is *tidal air*, which is inspired and expired in ordinary quiet breathing.

Vital capacity. The volume of air that can be made to pass into and out of the lungs by the most forcible inspiration and expiration is termed the *vital capacity of the lungs.* It is measured by means of a spirometer. In a normal man it is 4–5 litres and in a normal woman it is 3–4 litres. The vital capacity is reduced by diseases of the lungs, by heart disease (which causes congestion of the lungs) and by weakness of the muscles of respiration.

RATE AND CONTROL OF RESPIRATION

The mechanism of respiration is regulated and controlled by two principal factors, (*a*) the chemical, and (*b*) the nervous control. Certain factors stimulate the respiratory centre, which lies in the medulla oblongata, and when stimulated the centre generates impulses which are transmitted by spinal nerves to the muscles of respiration—the diaphragm and intercostals.

Nervous Control. The respiratory centre is an automatic centre in the medulla oblongata from which *efferent impulses* pass to the muscles of respiration. By means of some of the cervical nerve roots, impulses are conveyed to the diaphragm by the phrenic nerves; and at a lower level of the spinal cord, impulses pass from the thoracic region via the intercostal nerves to stimulate the intercostal muscles. These impulses cause rhythmical contraction of the diaphragm and intercostal muscles at the rate of about fifteen times per minute.

Afferent impulses stimulated by distension of the air sacs are carried by the vagus nerves to the respiratory centre in the medulla.

Chemical Control. It is this which is the ultimate factor in controlling and regulating the frequency, rate, and depth of the respiratory movements. The respiratory centre in the medulla is extremely sensitive to the reaction of the blood; the alkaline reserve of blood must be maintained (*see* page 170). Carbon dioxide is an acid product of metabolism, and this acid chemical substance stimulates the respiratory centre to send out nerve impulses which act on the muscles of respiration.

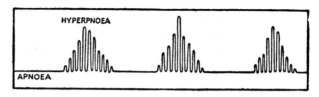

FIG. 16/7.—CHEYNE-STOKES BREATHING

An irregularity of respiration in which periods of deep breathing alternate with cessation of breathing, seen in the late stages of heart disease and other serious conditions such as uraemia.

Both controls, nervous and chemical, are essential; without either one of them man cannot continue to breathe. In cases of paralysis of the muscles of respiration (intercostals, and diaphragm), a pulmonary ventilator or some other means of continual artificial respiration is employed, because the chest must move in order that air may be carried in and out of the lungs.

Certain other factors will cause increase in the rate and depth of respiration. *Vigorous exercise*, by using up the oxygen in the muscles in order to provide the energy (work) needed, will give rise to a slight increase in the amount of carbon dioxide in the blood and result in fuller ventilation of the lungs.

Emotion, *pain*, and *fear*, for example, cause impulses to be registered which stimulate the respiratory centre and give rise to the sharp intake of air with which we are all familiar. *Afferent impulses from the skin* will produce a similar effect— when the body is plunged into cold water, or feels the first shock of a cold shower-bath, a sharp deep inspiration follows.

Voluntary control of the movements of respiration is possible but slight, as the movements are automatic. Any attempt to hold the breath for a longish time fails because of the discomfort caused by any increase above the normal amount of carbon dioxide in the blood.

The Rate of Respiration is slightly quicker in women than in men. In normal breathing expiration succeeds inspiration, and is followed by a slight pause. Inspiration—expiration—pause. In sick babies this order is sometimes reversed and the sequence becomes: inspiration—pause—expiration. This is described as *inverse breathing*.

Normal rate per minute:

In the newly born	40
At twelve months	30
From two to five years	24
In adults	10–20

Respiratory Movements. Two movements occur during respiration: (*a*) inspiration and (*b*) expiration.

Inspiration is an active process brought about by muscular action. Contraction of the diaphragm enlarges the chest cavity from above downwards, that is vertically. Elevation of the ribs and sternum, brought about by contraction of the intercostals, enlarges the chest cavity from side to side and from back to front. The elastic lungs expand to fill this increased space, and air is drawn into the air passages. The external intercostals are brought into play as accessory muscles only when inspiration becomes a conscious effort.

In *expiration* the air is forced out by relaxation of the muscles, and by the elastic recoil of the lungs. This is a passive process.

In *forced respiration* the movements of the chest are greatly increased. The muscles of the neck and shoulders help to raise the ribs and sternum. The muscles of the back and abdomen are also brought into action and the *alae nasi* of the nose alternately dilate and relax.

The need of the Body for Oxygen. In many conditions, including those already mentioned, oxygen may be ordered. People depend on oxygen for their lives; if completely suspended for more than four minutes, irreversible damage is caused to the brain and the patient usually dies. This emergency arises when, for example, a child at play pulls a Polythene bag over his head and face and is asphyxiated.•But if the supply of oxygen is only diminished, the patient becomes confused—he is suffering from *cerebral anoxia*. This happens to people working in a confined space, as in the holds of ships, in tanks and boilers; they use up the oxygen available and unless they are supplied with oxygen to breathe or are removed to normal air conditions, they will die of *anoxaemia*, shortened to anoxia. The alternative term is *hypoxaemia*, or hypoxia.

When there is insufficient oxygen in the blood, it loses its bright red colour, becoming bluish, and the patient's lips, ears and extremities are bluish in colour and he is *cyanosed.*

People attempting suicide by putting their heads into a coal gas oven are not only exposed to anoxia but they breathe the *poisonous carbon monoxide* which readily combines with the haemoglobin of the red cells, displacing the normal oxygen content. In these cases the lips are not bluish, but characteristically cherry red. The treatment needed is inhalation of and exposure to high concentrations of oxygen, containing the equivalent of up to five times the amount of oxygen in atmospheric air or five atmospheres.

Clinical Notes

Good room ventilation is essential, particularly where young people congregate, as in schools, offices and workshops, (*a*) in order to prevent the spread of respiratory infections, such as colds, influenza and bronchitis, or of any of the communicable diseases which may pass readily from one person to another, and (*b*) to avoid discomfort due to heat, humidity and stuffiness and maintain a 'fresh' atmosphere conducive to concentration on work or study.

(*See* note above, on the *need of the body for oxygen*).

Pulmonary ventilation, or the amount of air passing in and out of the lungs, may be too small if the breathing is weak because of damage to the brain and spinal cord, nerves, muscles or ribs, or if the breathing is obstructed by blockage of the air tubes as in asthma. Too small a pulmonary ventilation causes *anoxia* and accumulation of CO_2. *Treatment* is aimed at helping the breathing with artificial ventilation or by inhalations to relieve the obstruction to the bronchial air tubes. Disease of the lung tissue as in pneumonia does not cause a small pulmonary ventilation but does cause anoxia.

Dyspnoea, or difficult breathing, may be due to weakness of the nerves or muscles, damage to the ribs or pleural cavities, stiffness of the lungs due to pneumonia or pulmonary oedema in heart failure, or obstruction to the air tubes as in asthma or bronchitis. *Cyanosis* frequently accompanies these conditions.

In **lobar pneumonia** the areas affected are congested and the diffusion of oxygen impaired. The respiratory rate is increased in an attempt by the normal lung tissue to compensate for the failure of the congested parts.

In **bronchitis** as in pneumonia both ventilation and the diffusion of gases are impaired, swelling of the lining membrane obstructing air entering the lungs. *Chronic bronchitis* may be complicated by *emphysema* when air is retained in the lung tissue because the elastic tissue of the tiny air cells has degenerated and the air sacs remain permanently dilated with impairment of the membranous surface for the diffusion of gases. In *asthma* the air passages are narrowed and partially obstructed by muscular spasm. Expiration is particularly difficult. In *bronchiectasis* the bronchial tubes are dilated and often become infected.

Respiratory failure is failure of the respiratory function to keep the arterial oxygen and carbon dioxide content normal. There are two types: in the first, there is failure of pulmonary ventilation due to failure of the central nervous drive, as in overdosage by sedative drugs; failure of the peripheral nervous drive, as in poliomyelitis; failure of the chest bellows, as in extensive fracture of the ribs, or pneumothorax, or blockage of the larynx. In these conditions there is both lack of oxygen and excess of carbon dioxide.

In the second type of respiratory failure there is inadequacy of alveolar-capillary exchange, as in pneumonia or pulmonary oedema (page 159). In these conditions there is lack of oxygen but not an excess of carbon dioxide because, being more soluble, CO_2 can get out of the blood, even when the alveolar membrane is too thick for O_2 to get through.

Thoracic surgery is well established. **Thoracotomy** means opening the chest in order to operate on one of the organs within the thorax, such as heart or lungs, or on one of the structures in the mediastinum. *Thoracoplasty* is an operation for removal of certain ribs in order to permit the chest to fall in or collapse in order to immobilize a tuberculous lung. *Resection of rib* is performed to drain an empyema.

Operations on the lungs include *pneumonectomy*, when a lung is removed, and *lobectomy*, when one or more lobes are removed. A number of other operations include removal of a segment of lung tissue, *segmental resection*, or, if of a wedge of tissue, *wedge resection*.

Preparation for a thoracic operation is as for any major operation and includes routine chest X-rays, physiotherapy, and possibly diagnostic bronchoscopy. Any acute infection affecting the respiratory tract would delay operation until it had been treated adequately.

Operations on the oesophagus, for congenital anomalies, stricture, and for carcinomatous growths, may be performed through a thoracotomy incision.

Operation for the repair of an oesophageal **hiatus hernia** is one of the commonest operations in this group (*see* page 148).

After any thoracotomy an interpleural drainage tube is brought out through the chest wall to an underwater seal drainage bottle. This acts as a one-way valve mechanism, allowing the free drainage of air, blood or serous fluid from the chest. The remaining lung tissue is then free to re-expand fully to fill the thoracic cavity completely.

It is an important nursing observation to note that the drainage tube is patent and to record the quantity and type of fluid drained over the twenty-four-hour period. This becomes part of the fluid balance record which should be kept accurately from day to day.

Chapter 17

METABOLISM

Metabolism is the word used to indicate the chemical changes which take place in the body necessary for the fulfilment of its vital functions. Each cell is made up of protoplasm, which has the power to take in oxygen and other necessary substances, and to discard certain other properties as waste matter, including carbon dioxide, but between these changes taking place in the cells lies a large field of chemical activity upon which all the functions of the body intimately depend.

There is a continuous balance between the building up, *anabolism* of complex substances and tissues with the consumption of energy; and the breakdown, *catabolism* of complex substances with the liberation of energy. During growth or recovery from illness anabolism is predominant. During starvation or illness catabolism predominates.

Rate of Metabolism. *Basal Metabolism* is the term used to describe the sum total of metabolic activities with the body in a condition of physical and mental rest. In this state the minimum of oxygen will be required as the tissues are working at a minimum.

The basal metabolic rate is estimated on persons who are resting in bed, have not had any food or fluid during the night and who have not been disturbed. Either the intake of oxygen or the output of carbon dioxide is measured.

The principal factors which influence the rate of metabolism include: body size, age, sex, climate including the degree of heat, the type of clothing worn, and the nature of the work. It is obvious that the rate of metabolism will depend on the activity of the individual. It will be higher in a manual worker than in an office worker leading a more sedentary life.

The state of nervous tension is a most important factor as this will affect the rate of breathing and the rate and force of the action of the heart.

The basal metabolic rate in disease is influenced by some abnormality of the thyroid gland. *Over-activity* of the thyroid gland raises the metabolic rate, as in hyperthyroidism (*see* page 279). *Under-activity* of the gland lowers the metabolic rate, as in cretinism and myxoedema.

To compensate for heat lost and to maintain the necessary production of energy to be used as heat or work, man requires food. The energy value of food has been standardized and is expressed in kilojoules (kJ) which are provided by:

Protein, yielding . . .	17 kJ (4·1 kcal) per gram
Fats	38 kJ (9·3 kcal) per gram
Carbohydrates . .	17 kJ (4·1 kcal) per gram

Calories are required:
    To prevent loss of weight,
    To maintain the body temperature, and
    To provide for the functional activity of all cells, tissues, glands, and organs.

    A man doing heavy manual work
      requires .   .   14·7 megajoules (MJ) (3,500 kcal)
    A sedentary worker requires   .   10·5 MJ (2,500 kcal)
    A man at rest needs .   .   .   7·5 MJ (1,800 kcal)
    A patient in bed may need as little
      as .   .   .   .   .   .   5·0 MJ (1,200 kcal)
    Infants and growing children require proportionately more joules than adults, per kilogram of body weight.

The foods which provide heat and energy are carbohydrates, fats, and, under certain conditions, proteins. A résumé of the process of the metabolism of these three types of food is given below.

METABOLISM OF CARBOHYDRATE

As the result of digestion and absorption (*see* opposite), sugars and starches appear in the blood as glucose. The normal blood sugar level is 100 mg glucose per 100 ml of blood. Glucose is readily diffused into the tissue fluid and into cells and there is a uniform concentration of it in the body fluids.

Glucose is stored in the liver and skeletal muscles as gly-cogen; liver glycogen is reconverted to glucose as required by the body. This process requires the action of insulin (*see* page 282). Muscle glycogen is used during muscular activity and replenished from the blood sugar glucose as required.

In many illnesses additional joules are needed by the body and because carbohydrates are the most easily digested and assimilated type of food, the carbohydrate intake is increased rather than that of protein or fat.

Digestion

Ptyalin (salivary amylase) converts cooked starch into mal-tose.

Amylase (a group of enzymes) converts all starches into maltose.

Intestinal ferments (enzymes):

Invertase breaks down sucrose into glucose plus laevulose (fructose). *Maltase* breaks down maltose into glucose. *Lactase* breaks down lactose into glucose plus galactose. These breakdown products are all monosaccharides.

Absorption

The *monosaccharides* are absorbed into the blood—the percentage of blood sugar is maintained by the insulin control and liver activity.

In the tissues—carbohydrates are oxidized to provide heat and energy. Excess is stored as fat, increasing body weight.

During the process of combustion CO_2 is eliminated as a waste product.

The waste products which result from the burning up of carbohydrates in the tissues are excreted—

By the lungs: Water (H_2O) and Carbon Dioxide (CO_2)

From the skin: ,,

In the urine: ,,

METABOLISM OF FAT

Fat not immediately required after its absorption (*see* page 272) is stored in the fat depots of the body in adipose tissue. When needed it is withdrawn from these depots and in the liver is converted into glycerol and fatty acids, the form in which it can be utilized in the body.

When fat is metabolized by the liver, there is a residue of ketone substances which can only be used by the body to a limited extent. If they are produced by the liver faster than they can be used, they accumulate in the blood, causing the condition of *ketosis*. This happens in starvation when the body has nothing to use but the fat in its adipose tissue, in diabetes and on a diet which is too rich in fat and poor in carbohydrate.

Digestion. *Gastric lipase* produces slight hydrolysis of fat.
  Pancreatic lipase ⎱ break down fats into glycerin and fatty
  Intestinal lipase ⎰   acids.

Absorption of glycerin and fatty acids by the lacteals which are then passed to the thoracic duct, and enter the blood stream.

In the blood—fat is carried to every cell of the body.

The liver assists in the oxidation of fats and prepares fats for deposition in the tissues. The term *desaturation of fats* is sometimes used to describe this preparatory action of the liver on fats.

In the tissues—some of the fat is oxidized (in the presence of carbohydrates) to give heat and energy. Some of the fat is stored in the fat depots. (This stored fat contains vitamins A and D.)

The waste products which result from the combustion of fat in the tissues are excreted:
  By the lungs, water and carbon dioxide,
  By the skin, water, and
  By the kidneys, water.

METABOLISM OF PROTEIN

A great number of amino-acids are formed as the result of protein digestion and these form a pool (the amino-acid pool) from which the cells of the body draw the protein they need. Actually only nine of these amino-acids are essential for the growth and repair of body tissues. When diets contain an *excess of protein*, the surplus amino-acids are deaminated in the liver to remove the nitrogen, leaving only carbon, hydrogen and oxygen which can be used for the production of heat and energy. Conversely when protein intake is insufficient, as in *starvation*, not only are the carbohydrate and fat stores depleted, but there is a loss of body protein shown by wasting of the muscles. An example is seen in *kwashiorkor*, occurring principally in tropical and subtropical countries, when protein is very deficient in the diet (*see* Clinical Note, page 209).

Digestion. In the stomach:

  Pepsin (with HCl) converts proteins to peptones,
  Rennin produces casein from caseinogen, and
  Pepsin (with HCl) turns casein into peptones.

In the intestine:

  Trypsin reduces protein and peptone to polypeptides, and
  Erepsin further reduces polypeptides to amino-acids.

Absorption. Into the blood—the amino-acids bring nitrogen and sulphur to every cell in the body.

The body cells select the special amino-acids each cell needs for repair and growth.

The liver deaminates amino-acids and from this process urea is formed; carbon compounds are liberated for oxidation.

The waste products which result from the metabolism of protein in the tissues are—urea, uric acid, and creatinine. These substances are excreted in urine.

Protein is not stored in the body but excess is excreted, principally in urine.

Control of Metabolism. Consideration of the co-ordination of the activities of the different organs of the body will make it clear that some marvellous controlling mechanism is functioning in order to ensure that each cell does not function merely as a unit but as part of an organization—the body. The two most important controlling factors are:

The nervous system, central, and involuntary. An example of what happens when a muscle is deprived of its nerve supply is seen in infantile paralysis when the muscle wastes, the part ceases to function, and growth is retarded.

The endocrine organs. It is known that certain organs,

described as endocrine (*see* Chapter 18), produce substances of a chemical nature which control the well-being of the body, and in so doing effect changes in other organs. For example, when the secretion of the thyroid gland is diminished, metabolic activities are decreased; and conversely, when the secretion is increased or abnormal in character, metabolism is carried on at a greater rate as instanced by the rise in temperature and increased pulse rate characteristic of hyperthyroidism and the demand of the tissues for an increased supply of oxygen (*see* note on page 279).

Another point which must be considered in regard to the control and regulation of metabolism is the fact that increased activity of one organ may and often does lead to increase in the activity of other organs. For example, muscular activity results in better elimination of carbon dioxide; the presence of this gas in the blood stimulates respiratory activity; as a result there is greater intake of oxygen and the heart beats more forcibly in order to distribute the oxygen to the tissues, in this case to the muscles, where it is needed for the utilization of energy and the elimination of waste products.

THE MAINTENANCE OF BODY TEMPERATURE

The normal body temperature is 98·4°F or 36·9°C with a range of from 97° to 99°F or 36·1° to 37·2°C. The diurnal variation is about one degree Fahrenheit or half a degree Centigrade, the lowest level being reached in the early hours of the morning and the highest point between 5 and 7 p.m.

This normal temperature is maintained by an exact adjustment between heat produced and heat lost, and this is controlled by the heat-regulating centre in the hypothalamus, which is extremely sensitive to the temperature of the blood passing through it, acting like a thermostat.

Heat is produced by the metabolic activities in the skeletal muscles and liver. The glycogen stored in the liver is converted into usable glucose and oxidized, with the result that heat is produced. In order to maintain the normal production of heat the requisite amount of fuel food is necessary (*see* page 270). The metabolic activities (the rate of oxidation) must be adjusted to meet the varying demands made; for example, by active work or conditions of rest, the intake of food at meal-times and the periods between meals, the emotional reactions of the person, the external temperature, the clothing worn, and so on.

Overheating is usually due to a combination of a high external temperature, physical activity and inadequate sweating.

Heat loss is mainly effected by the functional activities of the skin (the importance of the skin in regulating the temperature of the body is mentioned on page 288). A certain amount of heat is lost by the evaporation of moisture from the lungs and by the excreta. To summarize, the organs concerned in producing heat and in losing heat are:

Heat Production	*Heat Loss*	Per cent
Fuel food, oxidized in all tissues	Skin—evaporation of sweat, radiation and conduction	75
	Lungs—evaporation of moisture	20
	Excreta	5
		——
		100

Heat loss is stimulated by vasodilatation in the skin and by sweating; *heat conservation* by vasoconstriction and diminished sweating. Conversely when the body temperature is lowered in prolonged vasoconstriction, due perhaps to exposure to cold or starvation, *shivering and shaking* may occur as the muscles contract to warm the body.

Clinical Notes

In bacterial invasion of the body, the temperature may be set higher, in *pyrexia*. This raised temperature, unless disordered (*see* below), acts as one of the defence mechanisms of the body, increasing the rate of metabolism which increases the need of the body for oxygen.

When the body temperature rises above 40°C it is described as *hyperpyrexia*; it may be so high that a special thermometer is needed to record it.

Hypothermia, for which a specially low-registering thermometer may be needed, occurs principally in infants and elderly people who do not have sufficient warm clothing or environmental heating, or who are deprived of warm food and drinks and who may, for one reason or another, be living on the border-line of starvation. (*See also* Metabolic Response in Illness, Chapter 13, page 210.)

Chapter 18

THE ENDOCRINE ORGANS

The endocrine organs or ductless glands are grouped together under this name because the secretion they make does not leave the glands by means of a duct, instead it is passed into the blood, circulating through the substance of the gland. The word *endocrine* comes from the Greek, and means 'internal secretion'; the active principle of an internal secretion is called *hormone*, from a Greek word meaning 'to excite'. Some of the endocrine organs produce a single hormone, others two hormones or more: the pituitary gland, for example, produces a number of hormones which control the activity of many of the other endocrine organs; for this reason the pituitary has been described as 'the master gland of the body'.

The endocrine organs are:

The *Pituitary*, anterior and posterior lobes,

The *Thyroid* and *Parathyroid* glands,

The *Adrenal* or *Suprarenal glands*, cortex and medulla, and

The *Thymus* gland and possibly also the *Pineal* body.

The formation of an internal secretion is an important function also of many other organs and glands, such as *insulin* from the Islets of Langerhans in the pancreas, *gastrin* in the stomach, *oestrogen* and *progesterone* in the ovaries and *testosterone* in the testes.

Knowledge of the function of these glands has been obtained by studying the effects of disease in them and this can usually be explained by the production of too much or too little of the necessary hormones.

PITUITARY GLAND

The Pituitary Gland (*see* Fig. 22/6, page 334) lies at the base of the skull, in the pituitary fossa of the sphenoid. It consists of two lobes, anterior and posterior, and an intermediate part, the pars intermedia. For the purpose of the study of its functions it is considered in two parts, the anterior and posterior lobes.

The Anterior Lobe of the pituitary produces a number of hormones which are instrumental in controlling the production of the secretion of all the other endocrine organs.

The *growth hormone* (*somatotropic hormone*) controls the growth of the body (*see* Clinical Notes, page 283).

The *Thyrotropic hormone* controls the activity of the thyroid gland in the production of *thyroxine*.

The *Adrenocorticotropic hormone* (ACTH) controls the activity of the adrenal glands in the production of cortisol from the cortex of the gland.

The *Gonadotrophic hormones* are:

The *follicle-stimulating hormone*, FSH, which stimulates the development of Graafian follicles in the ovary and the formation of spermatozoa in the testis.

The *luteinising* or *interstitial-cell-stimulating hormone*, LH, controls the secretion of the oestrogens and progesterone in the ovary and testosterone in the testis (*see* pages 312 and 318).

A third hormone, *luteotrophin* or *prolactin*, controls the secretion of milk, and maintains the existence of the corpus luteum during pregnancy.

Posterior Lobe Secretions. The posterior lobe of the pituitary secretes two hormones: *Antidiuretic hormone* (ADH), which regulates the amount of water passed by the kidneys, and the *oxytocic* hormone stimulating the contraction of the uterus during the birth of a baby and the release of milk during breast feeding.

THYROID GLAND

The thyroid gland consists of two lobes, placed one on each side of the trachea, and connected together by a strip of thyroid tissue, called the *isthmus of the thyroid*, which lies across, in front of, the trachea.

Structure. The thyroid gland is composed of numbers of vesicles lined with cubical epithelium (*see* Fig. 1/8, page 29), abundantly supplied with blood, and held together by connective tissue. These cells secrete a sticky fluid, *the colloid of the thyroid*, which contains an iodine compound; the active principle of this compound is a hormone *thyroxine*. This secretion fills the vesicles and from here passes to the blood stream either directly or through the lymphatics.

Function. The secretion of the thyroid is regulated by a hormone of the anterior lobe of the pituitary gland, the *thyrotropic hormone*.

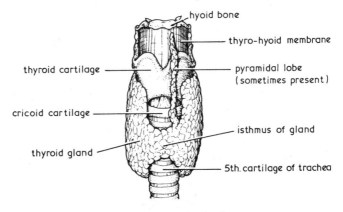

FIG. 18/1.—THE THYROID GLAND

The thyroid gland is intimately concerned with the metabolic activities regulating the chemistry of the tissues and is instrumental in stimulating oxidation processes and in regulating the consumption of oxygen and consequently the output of carbon dioxide.

Hyposecretion (*hypothyroidism*). Deficiency of the secretion of the gland at birth produces a condition known as *cretinism*, in which mental and physical growth are retarded. In adults deficiency of the secretion produces *myxoedema*; the general metabolic processes slow down, there is a tendency to gain weight, movements are lethargic; there is slowness of mind and speech, the skin becomes thickened and dry, and the hair falls out or gets thin. The temperature is subnormal, and the pulse slow.

Hypersecretion. In enlargement of the gland and increased secretion, *hyperthyroidism*, the symptoms are the opposite of those of myxoedema. The metabolic rate is raised and the body temperature may be higher than normal. The patient loses weight, is nervous and excitable, the pulse rate is raised, the cardiac output increased, and cardiovascular symptoms may include atrial fibrillation and heart failure.

In the condition known as Graves' disease or *exophthalmic goitre* the eyeballs protrude. This effect is due to over-activity of the thyroid hormone. It may not disappear when the disease is treated.

THE PARATHYROID GLANDS

The parathyroid glands are four small glands placed two on each side of the thyroid gland in the neck. The parathyroid secretion, *parathormone*, regulates calcium metabolism and controls the amount of calcium in blood and bone.

Hypoparathyroidism, in which there is deficiency of the blood calcium content, *hypocalcaemia*, causes a condition described

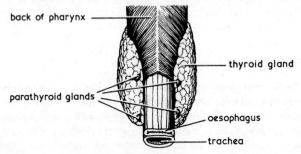

FIG. 18/2.—THE BACK OF THE OESOPHAGUS, SHOWING THE POSITION OF THE FOUR PARATHYROID GLANDS WHICH LIE *BEHIND* THE LOBES OF THE THYROID GLAND

as *tetany*, characterized by muscular twitchings and convulsions, particularly of the hands and feet, *carpedal spasm*; these symptoms are quickly relieved by the administration of calcium.

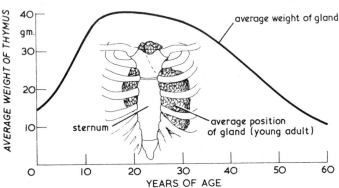

FIG. 18/3.—THE THYMUS GLAND

Hyperparathyroidism, over-activity of the glands, is usually associated with enlargement (tumour) of the glands. The balance of calcium distribution is disturbed, calcium is attracted out of the bones into the blood serum, with the result that a disease of bone, characterized by patches of rarefaction, occurs known as *osteitis fibrosa cystica* because bone cysts may be formed. The calcium may be deposited in the kidney, causing renal stones and kidney failure.

THE THYMUS GLAND

The thymus gland lies in the thorax about the level of the bifurcation of the trachea. It is pinkish-grey in colour and consists of two lobes. At birth the gland is quite small, weighing about 10 grams or a little more; it increases in size and at puberty weighs from 30 to 40 grams and then shrinks again. Its functions are unknown, but thought to be concerned with the production of antibodies.

THE ADRENAL GLANDS

The *Adrenal* or *Suprarenal glands* lie on the upper pole of each kidney. The adrenal glands consist of an outer yellowish part, the cortex, which produces *cortisol* (hydrocortisone), a close relation of cortisone, and an inner medullary portion producing both *adrenaline* (epinephrine) and *noradrenaline* (norepinephrine).

These substances are secreted under the control of the sympathetic nervous system. The secretion is increased in conditions of emotion such as anger and fear, and in states of asphyxia and starvation, and an increased output raises the blood pressure in order to combat the shock produced by these emergencies.

Noradrenaline raises the blood pressure by stimulating the muscular fibres in the walls of the blood vessels, causing them to contract. *Adrenaline* aids carbohydrate metabolism by increasing the output of glucose from the liver.

The important hormones secreted by the adrenal cortex are *hydrocortisone, aldosterone* and *corticosterone*, which are intimately concerned with metabolism, growth, renal function and muscle tone. These functions are essential to life.

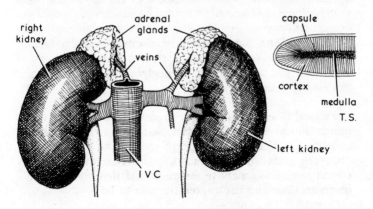

FIG. 18/4.—THE ADRENAL GLANDS

In *adrenal insufficiency* (Addison's disease) the patient becomes wasted and prostrated and gradually weaker, chiefly due to the fact that in the absence of this hormone, the kidneys fail to conserve sodium, which therefore is excreted in too large amounts. This disease is treated with cortisone.

The Islets of Langerhans of the pancreas (*see also* page 249) constitute an *endocrine organ* secreting *insulin*, the **anti-diabetic hormone,** given in the treatment of diabetes. Insulin is a protein which can be acted on by the protein digestive ferments and therefore is not given by mouth but by sub-cutaneous injection. Insulin controls and, when prescribed in a deficiency, such as in diabetes, restores the ability of the body cells to absorb and use glucose and fats.

Clinically, *deficiency* results in *hyperglycaemia*, a high blood sugar, loss of weight, fatigue and polyuria, with its accompanying thirst, hunger, dry skin, dry mouth and tongue. It also causes ketosis with acidosis and an increased rate of breathing.

The opposite condition, one of *hypoglycaemia*, a low blood sugar, may be produced by an *overdose of insulin*; or by a patient not eating (or perhaps vomiting) food taken after his injection of insulin when the excess in his blood may lead to *hypoglycaemic coma*.

Thus **coma** in a patient with diabetes may be due to lack of insulin **(diabetic coma),** which is treated with large doses of insulin, or too much insulin **(hypoglycaemic coma)** which is treated with glucose.

The Pineal Gland (*see* Fig. 22/6, page 334) is a small red body, similar in shape to a pine cone, situated near the corpus callosum. Its function is obscure.

Other glands which produce important internal secretions are the pancreas (*see* above and page 248), and the sex glands, which are described in Chapter 21.

Clinical Notes

Chemistry of hormones. The hormones of the pituitary gland, para-thyroids, adrenal medulla and the Islets of Langerhans, are digested in the stomach and when used therapeutically must be given by injection; those of the thyroid gland and the adrenal cortex, the corticoids, are not affected by the digestive enzymes and can be taken by mouth.

Pituitary glands. Pituitary dysfunction may be either of under (hypo-) secretion or of over (hyper-) secretion. Each leads to a well-recognized clinical syndrome. The following conditions are related to *dysfunction of the anterior lobe* of the pituitary. Hyposecretion before puberty leads to *dwarfism*. After puberty, in a previously normal individual a condition known as Sheehan's disease occurs. There are atrophic changes in the gonads, thyroid and adrenal glands. Hypersecretion before puberty gives rise to *gigantism* and after puberty to *acromegaly*, a condition in which the bones and also soft tissue organs thicken and coarsen. This particularly affects the hands, skull and jaw bones.

Posterior lobe. Failure of the posterior lobe to secrete sufficient ADH (antidiuretic hormone) leads to increased secretion of urine with accompanying thirst. This is known as *diabetes insipidus*. The *polyuria* can be most distressing, causing a patient to seek relief every few minutes. Injections of pitressin tannate give relief.

Thyroid gland. The secretion of the normal thyroid contains some iodine; and in countries where iodine is very deficient in the water supply, iodized salt or iodized sweets may be given. *Goitre* (a simple enlargement of the thyroid gland) is common in these countries.

The symptoms associated with hyposecretion (hypothyroidism) and hypersecretion (hyperthyroidism) are mentioned on pages 278–9.

Hyperthyroidism or *toxic goitre* must not be confused with the simple enlargement (non-toxic goitre) mentioned above, in which the gland is enlarged but its secretion is neither increased nor abnormal.

Hyperthyroidism may be treated with drugs which arrest the production of thyroxine in the gland, by operation or by radio-active iodine which is concentrated in the thyroid and bombards it with rays like X-rays.

In the treatment of hypothyroidism (*see* page 278) thyroid extract is administered to replace the deficiency, with the result that the symptoms are dramatically repressed.

Parathyroid glands. For hypoparathyroidism and hyperparathyroidism, *see* pages 279–80. *Osteomalacia* in adults and *rickets* in children are due to deficiency in bone calcium which these glands control. The cause is lack of vitamin D in the diet or failure to absorb vitamin D from the bowel. Vitamin D, being fat-soluble, is not absorbed when the digestion or absorption of fats is defective.

Adrenal or **Suprarenal glands.** Addison's disease of the glands is mentioned on page 281. This is a disease of *hypofunction*. *Hyperfunction* may be brought about by tumours of the adrenal glands producing Cushing's syndrome, in which there is obesity of the trunk, sparing the limbs, a 'moon' face, hypertension and biochemical evidence of defective carbohydrate and protein metabolism.

Pancreas. *Islets of Langerhans*—for Clinical Note *see* page 282.

Chapter 19

THE SKIN

The skin covers and protects the surface of the body and is continuous with the mucous membrane lining the cavities and orifices which open on to the surface. The skin has many functions; it contains the tactile nerve endings, helps to regulate the temperature and to control the loss of water from the body, and possesses some excretory, secretory, and absorptive properties. The skin is divided into two layers:

The Epidermis or Cuticle and
The Dermis or Corium.

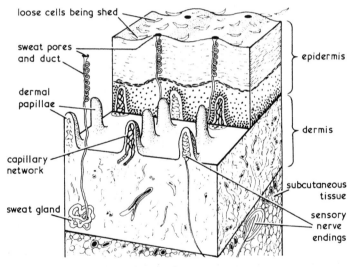

FIG. 19/1.—DIAGRAM OF SKIN SHOWING STRUCTURES

The Epidermis is composed of stratified epithelium and consists of a number of layers of cells arranged in two fairly well-defined zones: a horny zone and a germinal zone. The component parts of the epidermis can be distinguished

284

microscopically and Fig. 1/6 is here repeated for the convenience of the student.

Epidermal Layers. The *Horny zone* lies superficial. It is made up of the *three upper layers* of the cells of which the epidermis is composed.

Stratum corneum. Thin, flat, scale-like cells which are constantly being cast off.

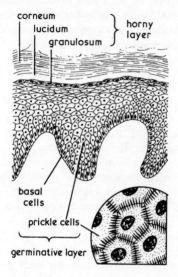

Fig. 19/2.—Microscopic Appearance of Epidermis

Stratum lucidum. Cells with an indistinct outline but no nuclei.

Stratum granulosum. A layer of well-defined cells containing nuclei and also granules—hence the term *granulosum.*

The *Germinal zone* lies beneath the horny zone and consists of two layers of well-formed epithelial cells:

Prickle cells, which are so-named because minute fibrils which connect one cell with another in this layer give individual cells the appearance of having prickles.

Basal cells. These are the cells from which new epidermal cells are constantly being produced. These cells are arranged in an orderly fashion, they are packed closely together and form the first layer or two of cells which rest on the papillae of the dermis.

The epidermis does not contain any blood vessels. The ducts of the sweat glands pass through it, and it accommodates the hairs. Epidermal cells line the hair follicles. The surface of the epidermis is marked by lines and ridges; these correspond to the papillae of the dermis which lie beneath. These lines vary; at the tips of the fingers and thumbs they form distinct patterns which differ in each individual. It is on this fact that the study of the fingerprints in criminology is based.

The Corium or Dermis is made up of fibrous and elastic connective tissue. The surface of the dermis is arranged in small papillae which contain loops of capillary blood vessels.

The nerve endings of the sensory nerves, the *tactile bodies*, lie in the dermis. The coiled tubes of numerous *sweat glands* lie in the deep parts of the dermis, and the ducts from these pass through the dermis and epidermis as spiral canals, to open on to the surface of the skin at minute depressions called pores. Some specially altered sweat glands are the *ceruminous glands* in the skin of the ear.

Sebaceous glands. These are small saccular glands found in the skin, they are flask-shaped and open into a hair follicle. These glands are most numerous in the scalp and face, around the nose, mouth, and ear, and do not occur at all in the skin of the palm of the hands and the soles of the feet. Both gland and duct are lined by epithelial cells. Changes in these cells result in the fatty secretion which is called *sebum*.

Appendages of the Skin. The *hairs* and *nails* and sebaceous glands are looked upon as appendages of the skin. Hairs and nails are modified epidermal cells. The *hair* grows from a hair follicle which is a deep recess in the epidermis (*see* Fig. 19/3).

The *hair follicle* is lined with epidermal cells and at the bottom of it is a papilla from which the hair grows. In health, when a hair drops out it is replaced by another hair grown

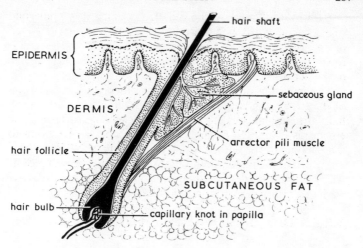

FIG. 19/3.—SKIN, SHOWING A HAIR FOLLICLE

A sebaceous gland is shown opening into the follicle.

from the same papilla. The root of the hair lies in the follicle. At its deepest extremity the hair is slightly thickened to form the hair bulb. This part fits over a vascular papilla and it is from soft cells in this region that the hair grows. The part which projects from the surface is the *hair-shaft*. The colour of the hair is due to the amount of pigment in the epidermis. Associated with the hair follicles are minute involuntary muscles, the *arrectores pilorum* or 'the raisers of the hairs', also *sebaceous glands* which secrete a fatty substance called *sebum*, which keeps the skin soft and smooth, and the hair glossy.

Nails. The nail is composed of modified skin. It lies on a *nail bed* in which the dermis is arranged in ridges instead of in papillae as in the skin. The nail bed is well supplied with nerves and is very vascular. The proximal part of the nail lies in a groove of the skin, the *nail groove*—it is thinnest in this region; and the white part, called the lunula, because of its shape, is the portion from which the nail grows forward. The *body of the nail* is the uncovered part, it is firmly attached to the nail bed.

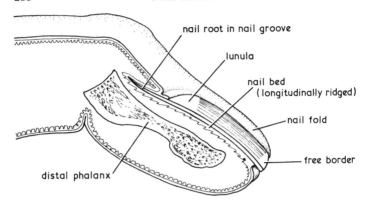

nail root in nail groove

lunula

nail bed
(longitudinally ridged)

nail fold

free border

distal phalanx

FIG. 19/4.—THE NAIL

The distal extremity of the nail is free— the *free border*—and at each side the nail is bounded by a fold of skin termed the *nail fold*.

THE FUNCTIONS OF THE SKIN

The Skin as a Heat-regulating Organ. The temperature of the body in man is constant, despite variations in his environment. It is maintained by an adjustment between heat loss and heat production, which is controlled by the heat-regulation centre. This becomes aware of any change in the body temperature, by the temperature of the blood passing through the medulla. The normal (deep) body temperature, that of viscera and brain, is 97° to 99·5°F (36° to 37·5°C). The skin temperature is slightly lower.

The vasomotor nerves control the state of the cutaneous arterioles by two actions, *vasodilatation* and *vasoconstriction*. In vasodilatation the arterioles are dilated, the skin gets hotter, and excess of heat is rapidly got rid of by radiation, by the increased activity of the sweat glands, and the subsequent evaporation of moisture from the surface of the body. In vaso-constriction the skin vessels are constricted, the skin becomes pale and cold, sweating is almost stopped, and the loss of heat is checked. By this control heat loss is increased or decreased according to the needs of the body.

The skin is the principal organ concerned in the loss of heat from the body. A considerable amount of heat is also lost by the lungs, and a little by the faeces and urine.

Heat is lost by the skin in various ways:

By *evaporation*, the amount of sweat formed depends on the amount of blood passing through the skin vessels,

By *radiation*, heat is given off to the surrounding air,

By *conduction*, the heat is transmitted to objects in contact, such as clothing, and

By *convection*, by movement of heated air in currents, the air in contact with the surface of the body is replaced by cooler air.

It is these factors which one takes into consideration when cooling an over-heated body, either by exposure of the skin to the surrounding moving air by fanning, sponging, or by immersion in cool water.

Sweat is an active secretion from the sweat glands under the control of the sympathetic nerves. It is essentially a salt solution whose concentration is about $\frac{1}{3}$ that of plasma. It is to be distinguished from *perspiration* or insensible water loss which is a simple diffusion of water through the skin. Perspiration loses about 500 ml of water per day. Sweating varies from 0 to 2,000 ml per day depending on the requirements of body temperature regulation.

The sweat glands are the main means by which *body temperature can be lowered*. A variable amount of water may be lost, about half a litre a day in a temperate climate, less in a cold and more in a hot climate. An environmental temperature higher than that of the body can be fairly comfortably supported if the air is dry; but humidity causes great discomfort as it prevents loss of body heat by evaporation.

The Skin as an Organ of Special Sense. The sensation of touch resulting from the stimulation of the nerve endings in the skin varies with the type of nerve ending stimulated. The sensations of heat, cold, and pain are all separate sensations. Certain spots exist in the skin called *sensory spots*; some of these are sensitive to cold, some to heat, and others to pain.

The sensations produced by deep pressure, and the sensation enabling a person to determine and judge the weight of an article, arise in the deeper structures such as the muscles and joints.

Storage. The skin and its underlying tissue acts as storage for water; the adipose tissue beneath the skin is one of the principal fat depots of the body.

Some of the Protective Properties of the Skin. The skin is relatively waterproof to the extent that it prevents loss of fluid from the tissues and it also prevents the passage of water into these tissues when, for example, the body is immersed in water. The epidermis prevents injury to the underlying structures and, covering as it does the sensory nerve endings in the dermis, it mitigates pain. When the epidermis is destroyed as in burns of the third degree, this protection being removed, every contact becomes painful, and exudation of fluid from the now exposed dermis causes serious loss of body fluid and electrolytes with the result that the patient is in danger of dehydration, that is water deprivation, and salt deprivation also which may cause serious illness.

Clinical Notes

The skin is so closely connected with the psychic mechanism of the individual that it acts as a mirror of the emotions: blushing with pleasure or shame, pallor and clamminess in fear. It is involved in a number of general infective conditions accompanied by rashes.

Skin diseases or *disorders* may be due to infective micro-organisms as in *impetigo*; to viruses as in *herpes*; to fungi as in *ringworm* and *athlete's foot*; to animal parasites as in *scabies* and *pediculosis*.

Many forms of *dermatitis* or *eczema* (inflammation of the skin) are due to an allergy to some food, drug or chemicals used externally or handled, such as powders, creams, oils, petrol, detergents and so on. Most of these are accompanied by erythema (redness) and urticaria (raised weals), conditions which often cause severe itching.

Urticaria may be produced by local contact in all subjects with sufficiently irritant substances, such as a wasp or nettle sting. It may also be produced locally in sensitive subjects by contact with the substances to which they are allergic, such as certain washing powders or cosmetics. It may also occur generally as a result of eating some food to which a subject is allergic.

Of other skin conditions such as *psoriasis* little is known of the cause. But all skin conditions are distressing, irritable, and require frequent attention for care and cure (when possible).

There are also malignant conditions—for example, *rodent ulcer* and *malignant melanoma*.

Chapter 20

THE URINARY SYSTEM

The urinary system consists of.
The Kidneys, which secrete urine,
The Ureters, to convey the urine from kidney to bladder,
The Bladder, which acts as a reservoir, and
The Urethra, for discharge of urine from the bladder.

The Kidneys lie on the posterior abdominal wall, mainly in the lumbar region, one on each side of the vertebral column, deeply embedded in fat, behind the peritoneum, and therefore outside the peritoneal cavity.

The position of the kidneys may be indicated, from behind, as extending from the level of the last thoracic vertebra to the third lumbar vertebra. The right kidney is slightly lower than the left, as the liver occupies considerable space on the right side.

Each kidney measures 10–13 cm (4–5 inches) in length, 6 cm (2½ inches) in breadth, and 2·5–4 cm (1–1½ inches) in thickness. An adult kidney weighs about 140 grams.

The kidneys are bean-shaped organs with the inner border or *hilum* directed towards the vertebral column. The outer border is convex. The kidney vessels enter and leave at the hilum. Each kidney is sur-

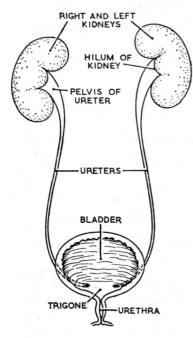

RIGHT AND LEFT KIDNEYS

HILUM OF KIDNEY

PELVIS OF URETER

URETERS

BLADDER

TRIGONE

URETHRA

Fig. 20/1.—The Organs which form the Urinary Tract

292

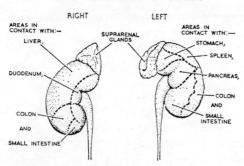

FIG. 20/2.—ANTERIOR SURFACE OF THE KIDNEYS, SHOWING THE POSITION OF THE
ADRENAL GLANDS AND THE RELATIONS OF THE KIDNEY

mounted by an *adrenal* (*suprarenal*) *gland*. The right kidney is
shorter and thicker than the left.

Structure of the Kidney. Each kidney is surrounded by a thin
capsule of fibrous tissue which invests it closely, forming a
smooth covering. Beneath this the kidney substance lies. It is
of a deep purple colour, and consists of an outer *cortical* part,
and an inner *medullary* part, which is made up of fifteen to
sixteen pyramid-shaped masses, the *pyramids of the kidney*. The
apices of these are directed towards the hilum, and open into
calyces which communicate with the *pelvis of the kidney* (*see*
Fig. 20/3).

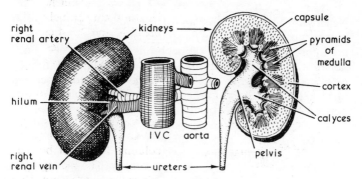

FIG. 20/3.—SHOWING THE OUTLINE, VESSELS, PELVIS AND GROSS STRUCTURE
OF THE KIDNEYS

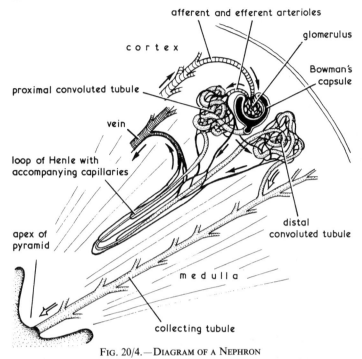

FIG. 20/4.—DIAGRAM OF A NEPHRON

The afferent arteriole is shown passing into a Malpighian capsule or glomerulus where it breaks up into capillaries. An efferent arteriole leaves the capsule and is shown breaking up into a second set of capillaries around the tubules.

Nephron. The *minute structure of the kidney* is composed of numbers of nephrons—the *functional units of the kidney*; there are approximately 1,000,000 nephrons in each kidney. Each nephron commences as a tuft of capillaries (Malpighian body or glomerulus), tightly packed into the expanded upper end of a uriniferous tubule or *nephron*. From it the tubule continues partly convoluted and partly straight. The first part of the tubule is convoluted and is known as the *first convoluted* or *proximal tubule*; this is followed by a loop, the *loop of Henle*; then the tubule again becomes convoluted, the *second convoluted* or *distal tubule*, which communicates with a collecting tubule that passes through the cortex and medulla to terminate at the apex of one of the pyramids.

Blood Vessels. In addition to the uriniferous tubules the kidney structure contains blood vessels. The *renal artery* brings blood from the abdominal aorta to the kidneys. Branches of this ramify in the kidney and break up into the afferent arterioles, each forming a knot of capillaries in one of the Malpighian bodies—these are the *glomeruli*. The efferent vessel then emerges as a small efferent arteriole, which breaks up to form a second capillary network around the uriniferous tubules. These capillaries eventually reunite to form the *renal vein*, which conveys the blood from the kidney to the inferior vena cava. The blood circulating through the kidney has therefore a double set of capillary vessels, the object being to retain the blood in the vicinity of the uriniferous tubules, upon which the function of the kidney depends.

RENAL FUNCTION

The Functions of the Kidney are the regulation of water balance; regulation of the concentration of the salts in the blood and of the reaction (acid-base balance) of the blood; and the excretion of waste products and any excess of salts.

The Secretion of the Urine and the Mechanism of Renal Function. The glomerulus is a filter. Every minute about 1 litre of blood, containing 500 ml of plasma, flows through all these glomeruli and about 100 ml (10 per cent) of it is filtered off. The plasma containing all the salts, glucose and other small substances is filtered. The cells and plasma proteins are too big to pass through the pores of the filter and stay behind in the blood stream.

The fluid that is filtered, the *glomerular filtrate*, then passes along the renal tubules and the cells absorb those substances which the body wants and leave behind those that are not wanted. By varying the amount that they absorb or leave behind the tubule cells can adjust the composition of the urine on one side of them and the blood on the other side. Normally, glucose is all reabsorbed; water is largely reabsorbed (*see* table p. 296), most of the waste products are excreted. In some special cases the tubules *add* substances to the urine. Thus the

secretion of urine consists of three factors:
  glomerular filtration;
  tubular reabsorption;
  tubular secretion.

By comparing the amount filtered by the glomeruli each day with the amount usually excreted in the urine we can see how selective the tubule cells are:

	Filtered	Excreted
Water	150 litres	$1\frac{1}{2}$ litres
Salt	700 grams	15 grams
Glucose	170 grams	0
Urea	50 grams	30 grams

The filtrate has by now reached the pelvis of the kidney and the ureter as urine.

Specific Gravity of the Urine. This depends on the amount of substances dissolved in or suspended in the urine. The specific gravity of the plasma (minus protein) is 1010. If the kidney is diluting the urine (e.g. after drinking water) the specific gravity is less than 1010. If the kidney is concentrating the urine (as it usually is) the specific gravity rises above 1010. The concentrating power of the kidney is measured by the highest specific gravity it can produce, which should be over 1025.

Tests of Renal Function. There are many of these but some simple ones include:

(1) *Testing the urine for protein* (albumin). If the glomeruli or tubules are damaged protein leaks into the urine.

(2) *Measuring the blood urea concentration.* If the kidneys are not excreting enough urea the blood urea rises above the normal value of 20–40 milligrams per 100 ml of blood. Glomerular filtration has to be reduced 50 per cent before there is a rise in the blood urea level, so that this is not a very sensitive test.

(3) *The concentration test.* No food is taken or water drunk for 12 hours to see how high the specific gravity rises (*see* above).

The Ureters are two ducts or tubes, one attached to each kidney, passing from it to the bladder. Each ureter is about the thickness of a goose-quill and 35–40 cm (14–16 inches) long. It consists of an outer fibrous covering, a middle muscular layer, and an inner mucous lining. The ureter commences as a dilatation at the hilum of the kidney, and passes down through the abdominal cavity into the pelvis to open obliquely into the posterior aspect of the urinary bladder.

THE URINARY BLADDER

The *bladder* acts as a reservoir for urine; it is a pear-shaped organ. It lies in the true pelvis in front of the other contents, and behind the symphysis pubis. In the infant it lies higher. The lowest part is fixed and is called the *base*, the upper part or *fundus* rises, the bladder becomes distended with urine. The apex lies forward beneath and behind the symphysis pubis.

The bladder consists of:

    An outer serous coat,
    A muscular coat,
    A sub-mucous coat, and
    A mucous lining, of transitional epithelium.

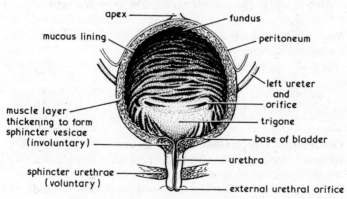

FIG. 20/5.—INTERIOR OF FEMALE BLADDER AND URETHRA SEEN FROM IN FRONT

Three vessels communicate with the bladder. The two ureters open obliquely into it at the base; their oblique direction prevents the regurgitation of urine into the ureters. The urethra opens out of the bladder inferiorly. The triangular area between the openings of the ureters and the urethra is the *trigone of the bladder*. In the female, the bladder lies between the symphysis pubis and the uterus and vagina. It is separated from the uterus by a fold of peritoneum—the utero-vesical pouch (*see* Fig. 21/1, page 304).

The urethra is a canal passing from the neck of the bladder to the external opening; it is lined with mucous membrane continuous with that lining the bladder. *The urinary meatus* is composed of circular muscle fibres, which form the *sphincter urethrae*. The female urethra is 2·5–3·5 cm (1–1½ inches) long, the male 17–23 cm (7–9 inches).

Micturition is the act of passing urine. As the urine is formed in the kidneys it passes along the ureters into the bladder. The desire to micturate is due to an increase of pressure in the bladder caused by the presence of urine there. This occurs when 170–230 ml (6–8 fl oz) have accumulated. Micturition is a reflex act which can be controlled and inhibited by the higher centres in man. The act is brought about by the contraction of the muscular coat of the bladder, and relaxation of the sphincter muscles. It may be assisted by contraction of the abdominal muscles which increases pressure in the abdominal cavity; the organs pressing upon the bladder assist in emptying it.

The bladder is controlled by the pelvic nerves, and sympathetic fibres from the hypogastric plexus.

Characteristics of normal urine. The *quantity* averages 1–2 litres daily in man, but varies greatly with the amount of fluids taken. It is also increased when excess protein is taken, in order to provide the fluid necessary to carry the urea in solution.

The *colour* is a clear, pale amber with no deposits, but a

light flocculent cloud of mucus may be seen floating in the specimen.

The *odour* is aromatic.

The *reaction* is slightly acid to litmus with an average *p*H of 6.

The *specific gravity* varies from 1010 to 1025.

Composition of normal urine. Urine is mainly water, urea and sodium chloride. In a man taking an average diet with 80 to 100 grams of protein in the 24 hours, the percentage of water and solids will be similar to the following:

> *Water* 96%
> *Solids* 4% (consisting of urea 2% and
> other metabolic products 2%)

Urea is one of the end products of protein metabolism. It is prepared from amino-acids, which are deaminated in the liver and reach the kidneys in the circulation, being excreted at the rate of 30 grams a day. The normal blood urea level is 30 mg per 100 ml of blood, but this depends on a normal intake of protein food and the function of the liver in the formation of urea.

Uric acid. The normal level of uric acid in the blood is 2 to 3 mg per 100 ml—1·5 to 2 g being excreted daily in the urine.

Creatinine is a waste product of creatine in muscle. *Other products of metabolism* include *purine* bodies, oxalates, phosphates, sulphates, and urates.

Electrolytes or salts such as sodium, potassium chloride are excreted to balance the amounts taken by mouth.

Clinical Notes

Medical diseases and disorders of the Urinary Tract.

Nephritis presents in several ways:

Acute nephritis with a rapid onset, raised temperature and pulse, scanty high-coloured urine containing albumin, *albuminuria*, and often blood, *haematuria*, giving it a smoky appearance.

Chronic nephritis may follow infective nephritis or pyelonephritis. There is *proteinuria* resulting in malaise, general weakness and anaemia. There may be hypertension, with danger of cerebral haemorrhage and congestive heart failure.

Nephrotic syndrome is a condition in which, although there is no failure of excretory function, the kidney loses large amounts of protein (particularly albumin); there is *gross proteinuria*, the plasma protein concentration falls and leads to oedema.

Pyelonephritis is inflammation of the tissue of the kidney and of the renal pelvis. It may be acute or chronic, occur in medical, surgical, or obstetrical conditions and is frequently associated with *cystitis* (*see* below). When acute the condition is painful, with a rise of temperature, shivering and vomiting. *Treatment* consists in giving bland fluids and keeping a careful fluid balance chart (*see* page 21). Chemotherapy is employed.

Chronic pyelonephritis is usually insidious and presents with hypertension and renal failure, rather than the symptoms of infection.

Cystitis, or inflammation of the bladder, may also be acute or chronic. In *acute cystitis* the passing of a very small quantity of urine at frequent intervals is painful, as it sets up *urethritis.*

Surgical conditions of the kidney. There may be *congenital* absence of one kidney or a horseshoe-shaped kidney may be present; *injuries* include bruising, laceration and rupture. The latter especially is accompanied by internal bleeding and shock.

Infections of the kidney include pyelitis, pyelonephritis and acute suppurative nephritis (distinct from acute nephritis). There may be tuberculous disease or malignant disease of the kidney.

The commonest surgical condition of the kidney is a *renal stone.* The formation of calculi (large stones) in the substance of the kidney can cause great damage; a stone in the ureter may block the outlet of urine from the kidney and cause *hydronephrosis* or dilatation of the renal pelvis. A stone passing down the ureter causes *renal colic*, which is excruciatingly painful.

Stone in the urinary bladder may have formed there or have passed into the bladder from the kidney. As the bladder contracts during micturition the stone is pressed against the sensitive trigone causing great pain. There is usually some haematuria, and urinary infection often accompanies these conditions.

Disorders of micturition include frequency, incontinence, nocturnal incontinence or enuresis, and dysuria, when pain and difficulty accompany the act. *Retention of urine* may be acute and painful or chronic and practically painless. The commonest causes are obstruction to the passage of urine by benign enlargement of the prostate gland or by stricture of the urethra, or by a calculus (stone).

Renal failure. *Acute renal failure* may be produced by acute nephritis, by renal toxins, or very commonly, by a period of low blood pressure which deprives the kidneys of their blood supply.

The urinary output is diminished, *oliguria*, to a few hundred ml a day, decreasing until there is complete suppression, *anuria*. The patient is extremely ill, his condition needs immediate treatment which will depend on the degree of renal deterioration and he requires expert nursing. When considered advisable, treatment is directed to restricting body fluids and electrolytes until renal function is restored. A careful fluid balance chart is kept.

Haemodialysis. Many patients, however, need *haemodialysis*, either *extracorporeal* by the *artificial kidney* or *peritoneal dialysis*. In the former the patient's blood is pumped through a Cellophane membrane rotating in a bath of dialysing fluid where waste products are removed, thus augmenting the function of the kidneys, and the blood is then pumped back into the patient's circulation.

Acute renal failure may be reversible or irreversible; there is no cure for irreversible failure but the condition can be relieved by haemodialysis at considered intervals.

Chronic renal failure is usually due to chronic nephritis, pyelonephritis, or malignant hypertension. There is an increased urine volume (*polyuria*) due to the inability of the kidneys to concentrate the urine, and uraemia.

Uraemia is a term used to describe the toxic condition mentioned above, due to the presence of renal waste products in the blood when on examination the quantity of urea present, which is not itself a toxic substance, is used to indicate the presence of other nitrogenous constituents which are toxic.

Chapter 21

THE ORGANS OF THE REPRODUCTIVE SYSTEM

The mode of development of the generative organs is interesting. The germ cells of the testis in the male and of the ovary in the female appear early in embryonic life. Sex therefore is determined from the very earliest days but sex characters cannot be recognized. It is a great and wonderful mystery how these reproductive cells are carried to the exact areas for which they are designated, the ovary and testis. They are developed in front of the kidney and are then carried in as columns of cells which eventually form the glands of reproduction consisting of germ cells and surrounding structures.

The *ovum* is the germ cell in the ovary and the *spermatozoon* the cell in the male. At adolescence these germ cells develop along with the changes which determine the sex qualities and characters of the male and female.

The organs of reproduction form what is known as the *genital tract* which is related to the urinary tract. In the male the two tracts are closely associated (*see* Fig. 21/10). In the female though the genital tract is in close relationship with the urinary tract they are not connected. The female genital tract communicates with the peritoneal cavity. The male tract does not do so; it is a closed tract. The female generative organs lie in the bony pelvis, the male organs lie mainly outside the pelvis.

PUBERTY

Puberty usually appears at 10 to 14 years and in girls is marked by the onset of menstruation—*the menarche*. The uterus and vagina enlarge; the breasts enlarge, with increase of fat, connective tissue and blood vessels. Later the secondary sexual characteristics appear; the curves of the body develop and adipose tissue rounds off the contours of her limbs, with the appearance of hair in the axilla and pubic region. The pelvis widens. Important changes take place as the girl matures mentally and emotionally through adolescence to womanhood.

In boys puberty is a little later. It is characterized by deepening of the voice, enlargement of the external genitalia, and the appearance of hair on the body and on the face.

THE MENOPAUSE

At the menopause, or climacteric period of a woman's life which occurs about 45 to 50 years, but may be earlier or later, menstruation ceases and is often accompanied by certain phenomena; vasomotor changes occur, with flushing and sweating: 'hot flushes'. The breast tissue often shrinks but it may be replaced by fat if there is a tendency to obesity. Senile changes take place in the ovaries, which become smaller, and their internal secretions are no longer formed.

THE PELVIC CAVITY

The *pelvic cavity* lies below and communicates with the abdominal cavity (*see* page 50). The *true pelvis* is the bony basin formed by the ischium and pubis which make up the sides and front, and the sacrum and coccyx which form the posterior boundary. The *brim of the pelvis* is formed by the promontory of the sacrum at the back, the ilio-pectineal lines at the sides, and the crest of the pubis in front (*see* page 86).

The *outlet of the pelvis* is bounded by the coccyx in the median plane *behind*, by the symphysis pubis *in front*, and the pubic arch, the ischium, and ligaments passing from the ischium to the sacrum on *each side*. This outlet is filled in by the structures forming the floor of the pelvis.

The Pelvic Floor. The structures which lie within the boundaries of the pelvic outlet form the floor of the pelvis. Two muscles, the levatores ani and coccygeus, act as a *pelvic diaphragm* (*see* Figs. 21/2 and 21/4).

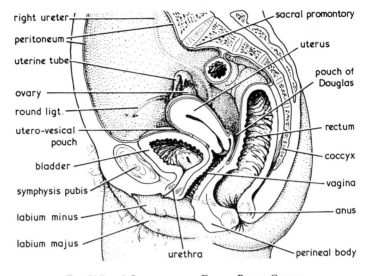

FIG. 21/1.—A SECTION OF THE FEMALE PELVIC CAVITY

The peritoneum covers the uterus, the recto-vaginal pouch (of Douglas) lies between it and the rectum
and the utero-vesical pouch between it and the bladder.

The *perineum* is the lowest part of the trunk. It is divided by a
line joining the two ischial tuberosities (upon which we sit) into
the *urogenital triangle* which is in front of this line and the
anal triangle which lies behind it. The central point is called
the *perineal body*; it is a strong fibrous muscular structure lying
in front of the anal canal, and in the female immediately
behind the vagina.

Contents of the Pelvis. (*See* Fig. 21/1.) The *urinary bladder* and
the ureters lie behind the symphysis pubis. The pelvic colon
lies in the left iliac fossa; the lowest or last part of the large
intestine, lies in the pelvic cavity.

The rectum, lying at the back of the cavity, follows the curve
of the sacrum.

Lymphatic vessels and *glands*, *nerves* from the *lumbo-
sacral plexus* (mostly destined for the lower limbs), branches of
vessels, from the internal iliac artery, and numerous *veins*, and

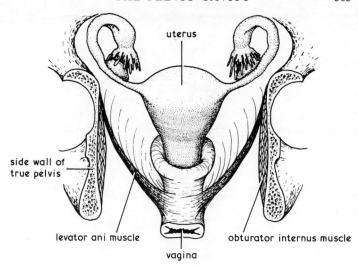

FIG. 21/2.—A DIAGRAM SHOWING HOW THE MUSCLES OF THE PELVIC FLOOR
FORM A SLING TO SUPPORT THE CONTENTS OF THE PELVIS

Compare this with Fig. 21/4 which shows how these muscles actually form the pelvic floor.

the *pelvic peritoneum* completes the contents of the pelvic cavity in the male. (*See also* page 320.)

The female pelvis also contains the uterus and its ligaments, the uterine tubes and the ovaries. (*See* Fig. 21/1.)

THE FEMALE ORGANS OF GENERATION

The organs of generation or the reproductive organs may be divided into the external organs and the internal organs.

The External Organs are collectively known as the *vulva*, and comprise the following parts:

The mons veneris, a pad of fat lying in front of the symphysis pubis. This area becomes covered with hair at puberty.

The labia majora are two thick folds which form the sides of the vulva. They are composed of skin and fat, and unstriped

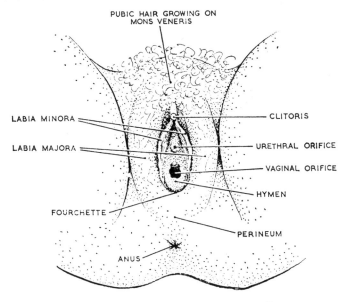

FIG. 21/3.—A DIAGRAM OF THE FEMALE EXTERNAL GENITALIA

muscular tissue, blood vessels, and nerves. The labia majora are about 7·5 cm (3 inches) long.

The *nymphae* or *labia minora* are two small folds of skin situated between the upper parts of the labia majora. The labia contain erectile tissue.

The clitoris is a small erectile body which corresponds with the penis of the male. It is situated anteriorly in the vestibule.

The *vestibule* is limited on either side by the labial folds and leads to the vagina. The urethra also opens into the vestibule in front of the vagina just behind the clitoris. The *greater vestibular* (Bartholin's) *glands* are situated just behind the labia majora on each side. These glands secrete mucus and their ducts open between the hymen and the labia minora. The *hymen* is a thin membranous diaphragm which is perforated centrally to allow the menstrual discharge to drain away. It is placed at the orifice of the vagina, thus separating the external and internal genitals.

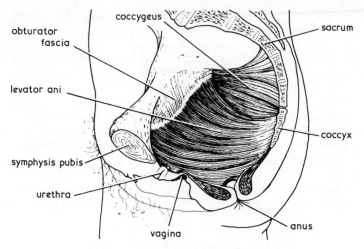

FIG. 21/4.—A LATERAL VIEW SHOWING THE ARRANGEMENT OF THE MUSCLES
WHICH FORM THE FLOOR OF THE PELVIS (*see also* FIG. 21/2)

An abnormal absence of the perforation just mentioned is a
rare condition, described as an *imperforate hymen*. The condition may not be noticed until the age when a girl should
menstruate; the discharge cannot escape, it collects in the
vagina, dilating it. Surgical incision of the hymen is undertaken, when menstruation can occur normally.

VAGINA

The vagina is a muscular tube lined with membrane comprised of a special type of stratified epithelium, well supplied
with blood vessels and nerves. The vagina extends from the
vestibule to the uterus. Its walls are normally in contact. It
surrounds the lower part of the cervix of the uterus (*see* Fig.
21/5), and rises higher behind than in front. The small recess in
front of and at the sides of the cervix are called the *anterior* and
lateral fornices, and the one behind the cervix is the *posterior
fornix* of the vagina.

The anterior surface of the vagina is in relation with the base
of the bladder and urethra, its posterior wall with the rectum

and the *recto-vaginal pouch* (of Douglas). The lower fourth of the vagina is in contact with the perineal body.

Structure. The vagina consists of three layers: an *inner layer* of mucous membrane characterized by ridges or rugae which give it the appearance of being covered with papillae (the mucous membrane of the vagina is of squamous stratified epithelial cells); the *outer layer*, a muscular coat of longitudinal and circular fibres; and between these coats is situated a layer of *erectile tissue* composed of areolar tissue, blood vessels and some unstriped muscular fibres.

The Internal Organs of Reproduction, which are situated in the pelvis, are the uterus, ovaries and uterine (Fallopian) tubes.

THE UTERUS

Structure. The uterus is a thick, muscular, pear-shaped organ situated in the pelvis, between the rectum behind and the bladder in front. The muscle is called the *myometrium*, and the

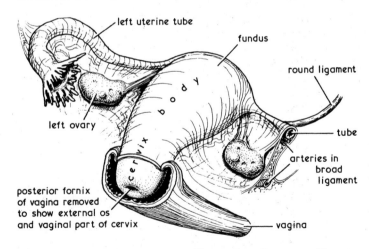

FIG. 21/5.—POSTERO-LATERAL VIEW OF UTERUS, LEFT OVARY AND UTERINE TUBE

mucous membrane which lines it inside is called the *endometrium*. Peritoneum covers most (not all) of the surface of the uterus, Fig. 21/1. It lies slightly anteflexed at the cervix and anteverted (turned forwards by rotation) with the fundus lying towards the bladder. It communicates below with the vagina and above the uterine tubes open into it. The broad ligaments are formed by two layers of peritoneum; ovaries and uterine tubes lie at the sides of the uterus. The blood supply is from the uterine and ovarian arteries. The uterus is 5–8 cm (2–3 inches) long and weighs 30–60 grams (1–2 oz). It is divided into the following three parts:

The fundus, a convex part above the openings of the uterine tubes.

The body of the uterus which extends from the fundus to the cervix, from which it is separated by the isthmus.

The lower narrow part of the uterus is called the *cervix*. The lumen of the cervix communicates with the cavity of the body of the uterus, via the *internal os* (os—mouth) and with that of the vagina via the *external os* (*see* Fig. 21/8, page 315).

Ligaments of the Uterus. The *round ligaments* are two bundles, one on each side, of connective and muscular tissue, containing blood vessels and covered by peritoneum, which pass from the upper angle of the uterus, forwards and outwards through the internal ring to the inguinal canal. Each round ligament is from 10 to 13 cm (4 to 5 inches) long.

Peritoneum dips down between the body of the uterus and the bladder in front, forming the *utero-vesicular pouch*, while behind it covers the body and cervix of the uterus and reaches down as far as the posterior fornix of the vagina before it passes on to the front of the rectum forming the *recto-vaginal pouch* (of Douglas).

Broad Ligaments. The peritoneum which covers the uterus in the mid-line of the body extends laterally on each side of the uterus as far as the side wall of the pelvis forming the broad ligaments. In the free edge of this broad ligament are the uterine tubes (*see* Fig. 21/5). The ovaries are attached to the posterior layer of the broad ligaments which is really the mesentery of the uterus and uterine tubes, and for this reason it contains the uterine blood and lymph vessels as well as those of the ovary.

Function of the Uterus. To retain the fertilized ovum during development. An ovum, when it is released from an ovary, is conveyed along a uterine tube to the uterus. (Fertilization of the ovum normally takes place in a uterine tube.) The endometrium has been prepared for the reception of the fertilized ovum (*see* page 313), which now becomes embedded in it. During pregnancy, which normally lasts about 40 weeks, the uterus increases in size, its walls become thinner but stronger and it rises out of the pelvis into the abdominal cavity as the fetus grows.

When term has been reached and labour commences, the uterus contracts rhythmically and expels the baby and placenta, and then returns to approximately its normal size by a process known as *involution*.

THE OVARIES

Structure. The ovaries are two almond-shaped glands placed one on each side of the uterus, below the uterine tubes, attached to the back of the broad ligament of the uterus. They contain a large number of immature ova, called *primary oocytes*, each one of which is surrounded by a cluster of nutritive follicle cells. At each menstrual cycle, one of these primitive ova begins to mature and quickly develops into a *vesicular ovarian follicle* (Graafian follicle).

As development of the Graafian follicle proceeds, changes take place in these cells, and fluid—the *liquor folliculi*—separates the cells of the *membrana granulosa* into layers. At this stage oestrogens are secreted. As the Graafian follicle approaches full development, or ripening as it is called, it lies near the surface of the ovary, gradually becoming more and more distended with fluid, until it projects as a cyst-like swelling from the surface of the ovary. Tension within the follicle causes it to rupture and the fluid and ovum escape via the peritoneal cavity into the funnel-shaped opening of the uterine tube (*see* Fig. 21/8).

Each month one follicle develops and one ovum is set free and extruded at about the middle (day 14) of the menstrual cycle. (*See also* page 313.)

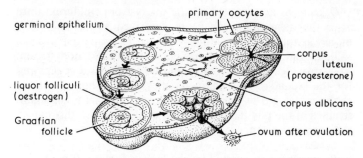

FIG. 21/6.—SECTION OF OVARY SHOWING MONTHLY CYCLE

Ovulation. Maturation of the Graafian follicle and liberation of the ovum is termed ovulation. When the Graafian follicle ruptures a little bleeding occurs, a clot is formed in the cavity of the follicle, and cells which have a yellow appearance grow into this clot from the wall of this follicle and form the *corpus luteum* or yellow body. Should the escaped ovum be fertilized the corpus luteum continues to grow for several months, becomes quite large and begins to atrophy at about 5 to 6 months.

If the ovum is not fertilized the corpus luteum persists for only 12 to 14 days, until just before the onset of the next menstrual period; it then atrophies and is replaced by scar tissue.

The Ovary has Three Functions:
(1) The production of *ova*, *see* ovulation above.
(2) The production of *oestrogens*. } Control of
(3) The production of *progesterone*. } menstruation.

The gonadotrophic hormones of the anterior pituitary control (via the blood stream) the production of hormones by the ovary itself. Follicle stimulating hormone (FSH) is essential for the early development of the Graafian follicle; and the pituitary also controls this growth by the luteinizing hormone (LH) and the secretion (luteotrophin) of the corpus luteum.

Oestrogens are secreted by the ovary from childhood until after the menopause. They are described as *follicular* hormones as they are constantly produced by numerous ovarian follicles and like all hormones circulate in the blood stream. They provide for the development of the female sex organs and for the secondary sex characteristics which bring about the changes in a girl at puberty, and are necessary for the maintenance of the physical and mental qualities which distinguish the normal woman.

Progesterone is secreted by the corpus luteum. It continues the work begun by the oestrogens on the endometrium and causes it to become thick, soft, and velvety, ready for the reception of a fertilized ovum. Progesterone inhibits menstruation.

The Onset of Menstruation is preceded by the degeneration of the corpus luteum which leads to a fall in the progesterone in the blood, but, during pregnancy menstruation does not occur. This is because the outermost cells of the *conceptus* (the chorion), as they burrow down into the endometrium, liberate a hormone (chorionic gonadotrophin) which acts on the corpus luteum and ensures the continuation of progesterone secretion. The ovarian secretions are thus regulated, not only by the pituitary but also by the chorion of the placenta which develops from the chorion later on in pregnancy (8–12 weeks).

The menstrual cycle consists of changes in ovary and uterus.
The *menstrual period* lasts about *five days*; during this period the surface epithelium is stripped off the lining of the *uterus* and bleeding occurs.
The *post-menstrual period* is a *stage of repair* and proliferation lasting about *nine days*, when the lining membrane is renewed. This stage is controlled by the oestrogens secreted by the ovaries, which in their turn are regulated by the FSH from

the pituitary (*see* page 277). Ovulation (*see* page 311) occurs at fourteen days and thereafter follows a further fourteen days known as the *secretory phase* which is under the control of progesterone secreted by the corpus luteum.

The endometrium becomes thick and soft ready for the implanting of a fertilized ovum, but if no ovum is fertilized congestion occurs in the capillaries ready for the menstrual period to follow.

The *periodicity of the menstrual cycle averages 28 days*, that is, 14 days in preparation for ovulation and a further 14 days. The endometrium is prepared for the arrival of a fertilized ovum, at about day 21. If only a non-fertilized ovum arrives in the uterus, then by day 28 the endometrium breaks down and menstruation occurs, and the cycle is repeated once more.

Fertilization is the result of the fusion of the male reproductive cell, the *spermatozoon*, with the *ovum* or egg cell which normally takes place in the uterine tube following sexual intercourse. A number of spermatozoa are deposited in the vagina (*see* Fig. 21/8). They pass through the uterus and find their way into the uterine tubes. Here resistance is met as the activity of the ciliated lining of the tube is directed to carrying the ovum along, from the other end of the tube, towards the uterus, but by the activity of their tails the spermatozoa are propelled. (One only is needed to effect fertilization.) Union of both cells is brought about by the spermatozoon penetrating the ovum. That union produces fertilization.

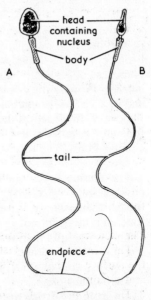

FIG. 21/7.—(A) SPERMATOZOON SHOWING COMPONENT PARTS (B) GIVES THE SIDE VIEW. THE GENETIC MATERIAL LIES IN THE HEAD

The fertilized ovum continues its journey down the tube towards the uterus and this takes about a week. As it progresses, it is sub-divided by cell cleavage into a number of small cells but the overall size remains constant. Arriving in the uterus, the outer cells of conceptus digest away part of the endometrium, the ovum sinks beneath the surface and implantation is said to have occurred. This normally takes place on the upper part of the body of the uterus near the opening of a uterine tube. But the conceptus may implant on any part of the endometrium and should it implant near the cervix the placenta will lie in front of the fetus.

It then becomes embedded in the wall of the uterus and development, which is giving rise to the formation of an entirely new complete human being, continues. Pregnancy has occurred.

UTERINE (FALLOPIAN) TUBES

The *uterine tubes* pass one on each side from the upper angles of the uterus outwards, in the upper margin of the broad ligament towards the sides of the pelvis. They are about 10 cm (4 inches) long, and at their uterine ends are narrow. They then enlarge, forming the *ampulla*, and finally bend downwards to end in a *fimbriated margin*. One of the fimbriae is attached to the ovary. In structure the uterine tubes are covered by peritoneum; beneath this lies the muscular coat of longitudinal and circular fibres. The tubes are lined by ciliated epithelial cells.

The uterine tubes open into the peritoneum, and thus a passage from the vagina, through the uterus and tubes into the peritoneal cavity, is formed, so that in the female the peritoneum is an open, not a closed, sac.

The ovaries and uterine tubes are supplied with blood by the ovarian arteries, and the nerve supply is derived from the hypogastric and ovarian plexuses.

The *normal function* of the uterine tubes is to convey the ovum from the ovary to the uterus. It also provides the sites where fertilization occurs.

But the journey of the ovum may be arrested at any point, and if such an ovum is fertilized, an *ectopic pregnancy* occurs.

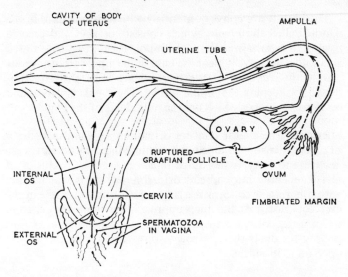

FIG. 21/8.—A DIAGRAM INDICATING THE ESCAPE OF AN OVUM FROM THE
OVARY

Spermatozoa are shown entering the vagina and arrows indicate the direction taken by the male
germ cell through the external os, cervix, internal os, cavity of the body of the uterus, and into
the uterine tube, in search of the ovum.
   Broken arrows indicate the escape of an ovum from the ovary and its passage into the
uterine tube, where fertilization normally takes place.

Unable to move onwards to the uterus, this ovum implants in
an abnormal site, usually within the uterine tube itself. Such
ectopic pregnancies usually terminate about 8–10 weeks later
by rupturing the tube. This requires emergency surgery.

THE MAMMARY GLANDS

The *mammary glands* or *breasts* are accessory to the female
reproductive organs and secrete the milk. (In the male these
glands are rudimentary.) The breasts lie in the superficial
fascia on the pectoral region between the sternum and axilla,
and extend from about the second or third, to the sixth or
seventh ribs. The weight and size of the breasts vary; they
enlarge at puberty and increase in size during pregnancy and
after delivery; they atrophy in old age.

The breasts are convex anteriorly with a prominence in the middle called the *nipple*, which consists of skin and erectile tissue and is dark in colour. The nipple is encircled by a tinted area called the *areola*. Near the base of the nipple are sebaceous glands, the *glands of Montgomery*, which secrete a fatty substance for keeping the nipple supple. The nipple is perforated by 15 to 20 orifices which are the milk ducts of the gland.

Structure. The breast consists of mammary gland substance or alveolar tissue arranged in *lobes* separated by fibrous, connective and fatty tissue. Each *lobule* consists of a cluster of alveoli opening into tubules or *lactiferous ducts* which unite with other ducts to form the larger ducts and terminate in the excretory ducts. As the ducts approach the nipple they expand to form reservoirs for the milk, these are called the *lactiferous sinuses*; the ducts then narrow to pass through the nipple and open on to its surface.

A considerable quantity of fat lies in the tissue on the surface of the breast, and also in between the lobes. Lymphatics are numerous. The *lymph vessels* commence as minute plexuses in the interlobular spaces of the gland tissue, unite and form larger vessels, which pass to the pectoral group of the axillary glands, the internal mammary, and the supraclavicular glands. The *blood supply* is derived from branches of the axillary, intercostal, and internal mammary arteries, and the *nerve supply* from the cutaneous nerves of the chest.

The Functional Activity of the Breasts. At birth the breasts often secrete milk, 'witches' milk', in the male as well as in the female.

In the female development changes occur at puberty when there is increase in the gland tissue. At the commencement of the menstrual life of a girl slight enlargement of the breasts takes place. This enlargement is due to the action of oestrogens and progesterone secreted by the ovaries (*see* page 311), and for a few days before each menstrual period the blood supply is increased; in some subjects this is more noticeable than in others, and gives rise to a sense of weight and slight congestion.

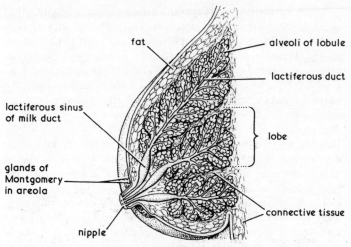

fat

alveoli of lobule

lactiferous duct

lactiferous sinus
of milk duct

lobe

glands of
Montgomery
in areola

connective tissue

nipple

FIG. 21/9.—THE RIGHT BREAST

Gradually the breasts become fully developed and the deposit of fat in the structure results in permanent enlargement, varying with the individual. At the menopause, that is at the other end of the menstrual life of a woman, when the ovaries gradually cease to function, the breast tissue shrinks (*see* page 303).

Lactation or the secretion of milk and its discharge from the breasts in suckling is the function of the breast. This may be considered in two phases:

The secretion of milk

Its discharge from the breast.

There is a little secretion in the breasts from the sixteenth week of pregnancy which keeps the ducts open and ready to function. After the birth of a baby a thin fluid, called *colostrum*, which is rich in protein, is secreted during the first 2–3 days; then a free flow of milk is established, gradually becoming mature milk. A hormone of the anterior lobe of the pituitary gland *prolactin* is important in stimulating the secretion of milk. The maintenance of this secretion is controlled by hormones from the anterior pituitary and the thyroid glands.

A *nursing mother* needs encouragement, especially with a first baby, to establish normal lactation. This does not depend only on the suckling efforts of the infant but also on a mechanism in the breast which by contraction expels milk from the alveoli into the ducts. Nervousness and other factors, such as doubts about breast feeding, can influence lactation, but with knowledge, practice and relaxation, this relationship between mother and baby can be happily established.

Disorders of the Breast. During lactation, in particular, a milk duct may become obstructed when a cyst which is called a *galactocele* may form. *Infection* may occur in any part of the breast, more usually during lactation.

The breast tissue may be the site of either simple or malignant tumour. The latter is fairly common in the breast and therefore any irregularity, swelling or contraction of the breast should at once be seen by a surgeon. In any infection or tumour of the breast the axillary lymphatic glands may be affected. Both infection and carcinoma are spread by permeation of the surrounding structures and by infiltration of the lymphatics.

THE GENITO-URINARY TRACT IN THE MALE

The urinary tract in the female is quite separate from the genital tract, but in the male it is not so separated. The *male urethra* is 17–23 cm (7–9 inches) long. It leaves the bladder and passes through the prostate gland, where it is known as the *prostatic urethra* which leads to the *membranous urethra*, which in turn leads to the *penile urethra*; curving at an angle of 90 degrees it passes through the perineum to the penis.

The *testes* are the male organs of generation where spermatozoa are formed and the male sex hormone, *testosterone*, produced. The testes develop in the abdominal cavity during fetal life, and descend through the right and left inguinal canals into the scrotum towards the end of pregnancy. There they lie obliquely suspended by the spermatic cords.

Testosterone, the male sex hormone, is secreted by the interstitial cells, i.e. the cells which lie in the interspaces between the seminiferous tubules of the testis under the stimulation of the luteinizing hormone (LH) of the pituitary. The secretion of testosterone increases markedly at puberty and is responsible for the development of the secondary sexual characteristics: growth of the beard; deepening of the voice; enlargement of the genitalia.

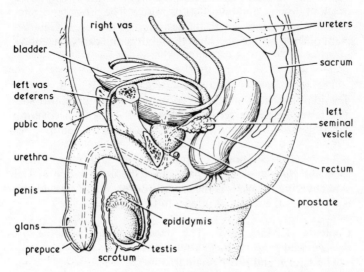

FIG. 21/10.—MALE REPRODUCTIVE ORGANS. THE GENITO-URINARY TRACT IN THE MALE

The *seminal vesicles* are paired tubular glands situated behind the neck of the bladder. Its duct joins with the *vasa deferentia*, Fig. 21/10, to form the common ejaculatory duct. The secretion of the seminal vesicle is an essential component of seminal fluid.

The *epididymis* is a small organ lying behind the testis and attached to it. It consists of a very long narrow tube which is extensively coiled up behind the testis. Through this tube the sperms pass from the testis into the vas deferens.

The *vas deferens* is a duct passing from the lower aspect of the epididymis. It ascends behind the testis, enters the spermatic cord and reaches the abdominal cavity through the inguinal canal, and finally passes into the pelvis.

The *prostate gland* is about the size of a large walnut; it lies below the bladder, surrounding the urethra, and is composed of glands, ducts, and involuntary muscle. The prostate secretes a fluid which mingles with the secretion of the testes. Enlargement of the prostate obstructs the urethra and causes retention of urine.

The *scrotum* is a pouch-like structure composed of skin devoid of subcutaneous fat; it contains a little muscular tissue. The testes lie in the scrotum, each testis lying in a covering called the *tunica vaginalis*, derived from the peritoneum.

The *penis* is composed of spongy tissue and is expanded to form the *glans penis* at the part where the urethra opens. The skin covering the penis is the *prepuce* or *foreskin*. *Circumcision* is the removal entirely or in part of the prepuce.

Contents of Male Pelvis. The *urinary bladder* with the *vas deferens* and *prostate gland* attached.

The *rectum* and *pelvic peritoneum*.

Lymphatic glands and *vessels, sacral nerves, arteries* and *veins*.

REPRODUCTION

Reproduction is brought about by the activity of the male and female essential sex organs—the testes and ovaries, producing male and female sex cells—sperms and ova.

The testes and ovaries are sometimes called *gonads*— male gonads and female gonads. These organs also produce hormones which give rise to the sexual development of man and woman and their sex characteristics. The production of these hormones is controlled by the gonadotropic hormones of the pituitary gland.

By the union of the two sex cells, male and female, human life is perpetuated. To enable this union of the reproductive cells to take place certain accessory male and female reproductive organs are required.

The **male accessory organs** are the *epididymis*—small tubules where the sex cells or sperms are stored in the testes—and the *vas deferens* by means of which semen is conveyed to the first part of the urethra, and the *penis*.

The urethra passing through the penis has two functions—the voiding of urine and the ejection of semen. The penis contains erectile tissue which enables it to become rigid and erect.

The **female accessory sex organs** are the *uterine tubes* through which the fertilized ovum passes in to the *uterus* to become embedded in the lining membrane thickened to receive it, the *cervix* and the *vagina* where, in the act of intercourse, semen is deposited.

Sex Determination depends on sex chromosomes. The normal number of chromosomes in man is 44 and in addition 2 sex chromosomes, making 46; a child receives 23 chromosomes from each parent. He receives 22 pairs of *autosomes*, which are the ordinary chromosome as distinguished from sex chromosomes. There are two sex chromosomes, X or Y. The sex is determined by the father of the child, because only the sperm carries the Y chromosome. The ovum contains 22 ordinary chromosomes and one X (sex) chromosome.

Thus 44 plus XX (two sex chromosomes), one X from mother and one X from father, *produces a female*.

But 44 plus XY, the X from mother and a Y sex chromosome from father, *produces a male*; the sex is determined by the father of the child, according to which of the 2 sex chromosomes (X or Y) he contributes.

Development of the Fetus. The term **conceptus** may be employed to describe development at any stage.

After implantation, already mentioned (*see* page 314), the

conceptus becomes embedded in the uterine endometrium, deriving nutrition from the maternal blood. It is during the first 10 weeks of this period, when the organs are actually being differentiated, that the embryo is more likely to suffer injury from outside effects, such as drugs (e.g. thalidomide) or infective agents, e.g. rubella (German measles), and the child be born later with an abnormality.

The **embryo** is enclosed within two membranes, the inner *amnion* and an outer *chorion* which constitute a bag of membranes or the *amniotic sac*. This is filled with a fluid, *liquor amnii*, which, by exerting equal pressure in every direction, serves to protect the fetus and also allows for the free movement and uniform growth of the fetus (the descriptive term usually employed after the 12th week).

During the first 8 weeks of development, the chorion is directly in contact with maternal blood. The surface area is increased by the development of villi (much as the lining membrane of the small intestine is covered).

About the 8th week the bag of membranes expands so as to obliterate the cavity of the uterus; the chorionic villi towards the uterine endometrium continue development and form the placenta (*see* page 323). Over the rest of the chorion the villi disappear. As the fluid in the bag of membranes increases, the fetus with umbilical cord attached moves buoyantly in what is virtually a state of weightlessness.

Certain milestones in the *development of the fetus* may be mentioned. At about 12 weeks the uterus rises out of the bony basin of the pelvis and can be palpated at the level of the symphysis pubis. Formation of all the organs is now complete. The remaining months are taken up by growth. By the 17th week fetal movements may be noticeable by the mother and at about 20 weeks fetal parts can be distinguished by abdominal palpation. At 28 weeks the fetus is *legally viable*, that is, capable of living a separate existence from the mother, should premature birth occur. In fact, many babies born before the 28th week survive and develop normally. The baby is normally ready to be born at 40 weeks when the pregnancy is said to have reached *term*, and now generally weighs 2·7–3·6 kg (6–8 lb) and measures about 50 cm (20 inches) in length.

The **placenta** is formed by about the 8th week of pregnancy (*see* page 322) from that part of the conceptus which is in direct contact with the uterine endometrium, to which it remains firmly attached until after the birth of the baby. Its functions are: to provide the fetus with nourishment derived from the maternal blood; to act as the 'fetal lung' by providing for the oxygenation of the fetal blood and the removal from the fetus of waste products.

The placenta also acts as a barrier in preventing certain micro-organisms of disease reaching the fetus.

Drugs. Most drugs can cross the placenta; any cerebral depressant drugs, such as morphine, barbiturates and general anaesthetics given to a mother during labour can depress the breathing in a newborn baby. Great consideration is given before prescribing any drugs for a pregnant woman; the *thalidomide disaster* which caused abnormalities in growth in the early weeks of pregnancy is well known.

The placenta helps the ovaries (*see* page 311) in the production of hormones necessary for the continuation of pregnancy, and these also play an important part in relation to lactation by stimulating the development of glandular breast tissue and its ducts.

The **Umbilical Cord** is a flexible structure connecting the fetus at the umbilicus to the placenta; it contains the vessels which carry blood to and fro between the fetus and the placenta (*see* fetal circulation, below). When the baby is born, the umbilical cord is ligatured and divided and thus all connection with the placenta is severed. The stump of the cord dries up and separates from the baby a few days afterwards, leaving the *umbilicus* or 'navel' marking its site of attachment, which is a 'scar'.

The Fetal Circulation. Points which should be kept in mind when studying the fetal circulation include:

(1) As the fetus gets its oxygen and food supply from the placenta, the whole of the fetal blood has to pass through it.

(2) All the blood is mixed, so that one cannot accurately speak of 'pure' and 'impure' blood, though these terms are used. It is more correctly described as *re-oxygenated* blood from the placenta but this never contains the oxygen saturation of 95 per cent to 100 per cent arterial blood in the adult and the blood is *de-oxygenated* when it leaves the fetus to re-enter the placenta. Fetal blood contains about 80 per cent oxygen saturation.

(3) The *functions of the lungs* are carried out by the placenta. *In utero* the fetus has no pulmonary circulation as in the adult circulation; a limited supply of blood reaches the lungs, sufficient only for their nutrition and growth.

(4) The fetal alimentary tract is also functionless, as the placenta supplies nourishment to and removes waste products from the fetus.

Therefore the fetus *in utero* has a circulation which differs considerably from that of post-natal life.

Re-oxygenated blood leaves the placenta by a single umbilical vein; the umbilical vein passes in the umbilical cord to the umbilicus and then a small vein passes to the porta hepatis. Hardly any of this blood passes into the liver because the umbilical vein is connected directly to the inferior vena cava by a large vessel, called the *ductus venosus*, a fetal structure. Once in the inferior vena cava (1) the blood travels *upwards* and reaches the right atrium (2). Most of the blood, instead of passing into the right ventricle (as one would expect from knowledge of the adult circulation) passes instead into the left atrium, through a temporary or fetal opening in the interatrial septum, called the *foramen ovale* (3). Having reached the left atrium (4) it passes through the mitral valve into the left ventricle (5). Contraction of the left ventricle pushes the blood into the ascending aorta (6), where most of it is distributed to the heart itself, to the brain and to the upper limbs. The blood remaining in the aortic arch (7) passes on into the descending thoracic abdominal aorta, and will be mentioned on p. 326.

Having circulated in the brain and upper limbs the blood returns towards the heart in the superior vena cava (8) and reaches the right atrium. Keeping on its downward course in

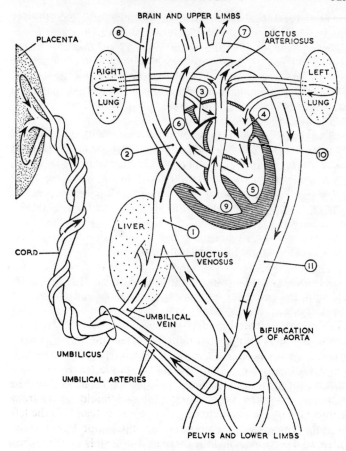

PLACENTA

RIGHT LUNG

LEFT LUNG

BRAIN AND UPPER LIMBS

DUCTUS ARTERIOSUS

CORD

LIVER

DUCTUS VENOSUS

UMBILICAL VEIN

UMBILICUS

UMBILICAL ARTERIES

BIFURCATION OF AORTA

PELVIS AND LOWER LIMBS

FIG. 21/11.—FETAL CIRCULATION

The *route taken by fetal circulation* can be followed by reference to the text where the figures correspond to those on the diagram; the *temporary fetal structures* are not numbered but printed in full on the diagram: these are the *umbilical cord, vein and arteries*, the *ductus venosus* and the *ductus arteriosus*.

the right atrium it then passes through the tricuspid opening into the right ventricle (9). From here it is pumped into the pulmonary artery (10). A knowledge of adult anatomy would suggest that the blood should then be distributed to the lungs.

In fact, the lungs are inactive in the fetus and receive very little blood. Most of the blood in the pulmonary artery is diverted straight into the aorta via a large muscular artery called the *ductus arteriosus* which joins the aorta near the end of the aortic arch. The descending thoracic aorta (11) therefore contains mostly reduced blood which has reached it via the ductus arteriosus and a smaller quantity of oxygenated blood already mentioned which reaches it from the aortic arch.

The blood in the aorta is then distributed to the abdominal viscera via the lower branches of the aorta. But in the fetus *most of the blood* which reaches the bifurcation of the aorta travels not to the pelvic viscera and lower limbs, as adult anatomy would suggest (yet these structures receive an adequate supply), but by means of a pair of large umbilical arteries to the fetus's umbilicus, then along the umbilical cord back to the placenta, where exchanges take place with the maternal blood across the placental barrier.

The fetus makes its own blood and so long as the placental barrier remains complete, there is no mixing. Having passed through the capillaries of the placenta the blood passes back to the fetus once more.

Certain changes take place at birth. The *valve-like flap which guards the foramen ovale* closes and the right and left atria are permanently separated. The *ductus arteriosus* also contracts down and fibroses. These two openings, normal during fetal life, may persist and give rise to abnormalities after the child is born.

When the umbilical cord is cut and ligatured, blood ceases to flow in the *umbilical arteries and vein* and in the *ductus venosus*. All these structures close down and are replaced by strands of fibrous tissue. The *ligamentum teres* of the liver is in fact a remnant of the umbilical vein.

Chapter 22

THE CEREBROSPINAL NERVOUS SYSTEM

The *Nervous System* is divided for description into two main parts: (1) the Central or Cerebrospinal system and (2) the Autonomic, which includes the Sympathetic and Parasympathetic Nervous Systems. (*See* Chapter 23.)

The Cerebrospinal Nervous System. This consists of the *brain* and *spinal cord* and the nerves given off from these, the *peripheral nerves*. Nervous tissue forms one of the four groups of the elementary tissues of the body.

Nerve cells massed together form what is called the grey matter of this system, as is found in the cortex of the brain, and in the inner part of the spinal cord.

Nerve fibres or axons form the white matter. This difference in colour is due to the axons or conducting fibres being covered by a sheath of fatty matter, which serves to protect, nourish, and insulate the nerve fibres from each other (*see* Fig. 22/2).

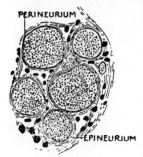

FIG. 22/1.—SECTION OF A NERVE TRUNK SHOWING SEVERAL FASCICULI SURROUNDED BY PERINEURIUM AND HELD TOGETHER BY EPINEURIUM. THE AREAS STAINED BLACK ARE FAT CELLS

A nerve cell with its axon and other processes constitutes a *neurone*. In the formation of a nerve trunk the nerve fibres are arranged in bundles called *fasciculi*.

A **nerve fibre** possesses the power of *conductivity* and *excitability*. It is capable of receiving and responding to stimuli from some outside agent— for example, the *stimulus* may be mechanical, electrical, chemical, or psychical; this gives rise to an *impulse* which is conducted along the nerve fibres. A nerve impulse is always conducted along a dendron to a cell, and from cell to axon.

327

This is the *law of forward conduction*. An impulse may be passed along a series of neurones in this way.

A **motor impulse** generated in one of pyramidal cells of the motor area of the cortex travels along the *axon* or nerve

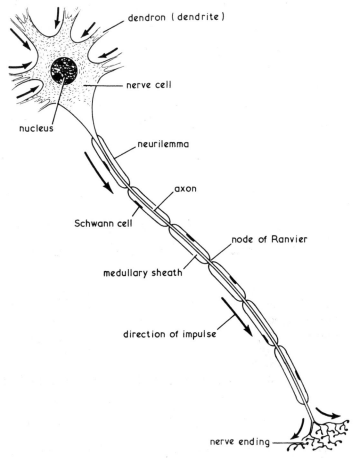

FIG. 22/2.—A MEDULLATED NERVE (DIAGRAMMATIC). THE AXON OR AXIS CYLINDER RUNS FROM THE CELL TO THE END OF THE NERVE FIBRE. IT IS SURROUNDED BY A FATTY SHEATH—THE MEDULLARY SHEATH—WHICH IS INTERRUPTED BY THE NODES OF RANVIER

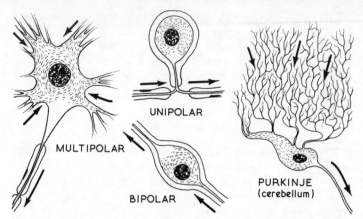

FIG. 22/3.—EXAMPLES OF NERVE CELLS. ARROWS INDICATE THE DIRECTION IN WHICH THE NERVE IMPULSE ALWAYS TRAVELS—FROM DENDRON TO CELL, AND FROM CELL TO AXON

fibre, which, passing down the spinal cord, lies in the white matter; the axon arborizes with the dendrites of motor nerve cells in the anterior horn of the spinal cord. The impulse then passes to the axons of these cells, which form the motor fibres of the anterior root of a spinal nerve, and is conveyed to terminate in a muscle.

Sensory impulses received by the nerve endings in the skin, travel by nerve fibres (*dendrons*) to the sensory cells in the posterior root ganglion, and thence by the axons of these cells into the spinal cord and ascend to a nucleus in the medulla, thence to be relayed to the brain. (*See* sensory nerve pathway, page 349.) Nerve fibres travelling to and from different parts of the brain are grouped together in definite tracts in the spinal cord.

Nerve trunks formed by the cerebrospinal nerves are of three varieties:

(1) *Motor or efferent nerves* carrying impulses from the brain and cord to the periphery (*see* motor pathway, page 348).

(2) *Sensory or afferent nerves* carrying impulses from the periphery to the brain (*see* sensory pathway, page 349).

(3) *Mixed nerve trunks* containing both motor and sensory fibres, thus carrying impulses in both directions. Most of the nerves are of this last variety.

In addition there are certain nerve fibres which link up different nerve centres in the brain and cord. These are called *associated or commissural nerve fibres.*

THE MENINGES

The brain and spinal cord are surrounded by the *meninges*, which protect the delicate nerve structure, carry the blood vessels to it, and, by the secretion of a fluid, the *cerebrospinal fluid* (*see* page 331) minimize any blow or concussion. The meninges are in three layers.

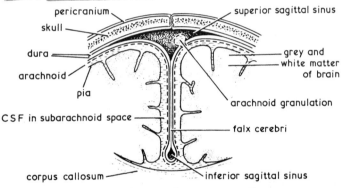

FIG. 22/4.—DIAGRAM OF A CORONAL SECTION OF THE CEREBRUM, SHOWING THE RELATIONS OF THE MENINGES TO THE SURFACE OF THE BRAIN

The pia mater dips into the fissures of the brain and cord and by this close contact supplies these structures with blood.

The **arachnoid** is a fine membrane separating the pia mater from the dura mater.

The **dura mater** is dense and tough; it consists of two layers; one outer layer lines the skull, an inner layer is united with it, except where the venous sinuses are formed and where the dura mater forms the following partitions:—The *falx cerebri* lies between the two cerebral hemispheres; its upper border

forms the superior longitudinal or sagittal sinus receiving
venous blood from the brain and its lower margin, the inferior
longitudinal or inferior sagittal sinus, drains the falx cerebri.
The *tentorium cerebelli* separates the cerebellum from the cerebrum. (Reference to Fig. 11/12, page 190 indicates the positions
of these sinuses and the partitions of the dura mater.)

The *diaphragma sellae*, a ring-shaped fold of dura mater,
covers the sella turcica, a depression on the sphenoid bone,
containing the *hypophysis cerebri*, or pituitary gland.

In **meningitis** there is inflammation of the meninges, characterized by increase in the amount and alteration of the composition of the cerebrospinal fluid (C.S.F.). The infection
may be bacterial or viral; diagnosis is made by examination
of the cerebrospinal fluid obtained by lumbar puncture (*see*
page 333).

The **Ventricular System** consists of several interconnected
cavities within the brain into which the cerebrospinal fluid is
secreted by the choroid plexuses. These *choroid plexuses* are
composed of a network of minute capillary blood vessels and
covered with pia mater which project into the ventricles and
secrete cerebrospinal fluid.

The *two lateral ventricles* lie one in each cerebral hemisphere
and are connected with the third ventricle which lies in the
mid-line between the thalami. The *third ventricle* is connected
by a narrow channel, the *cerebral aqueduct*, with the *fourth
ventricle* which lies between the cerebellum and the pons and
medulla. Openings in the roof of the fourth ventricle allow the
cerebrospinal fluid to pass into the sub-arachnoid space surrounding the whole of the brain and the spinal cord.

The **cerebrospinal fluid** is a secretion produced by the
choroid plexuses (*see* Fig. 22/5). It is a clear alkaline fluid
resembling plasma. The *pressure* is 60 to 140 mm water.

The *circulation of cerebrospinal fluid*. The fluid is secreted
by the choroid plexuses into the ventricles which lie within the

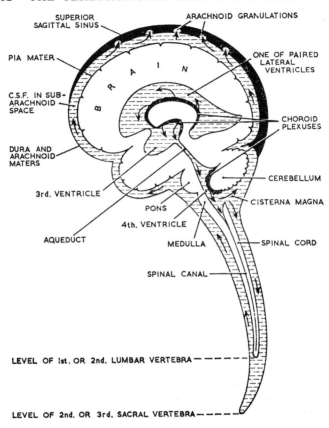

FIG. 22/5.—DIAGRAM INDICATING THE SITUATION OF THE SPACES CONTAINING
FLUID—SUB-ARACHNOID SPACE, VENTRICLES AND SPINAL CANAL—WHICH
LIES WITHIN AND AROUND THE BRAIN AND SPINAL CORD

brain; from openings in the fourth ventricle it passes into the
central canal of the spinal cord, and from the ventricles also
into the sub-arachnoid space. The fluid is now able to pass in
this space over the whole surface of the brain and spinal cord
until it is returned to the venous circulation by the arachnoid
granulations in the superior sagittal sinus (*see* Fig. 22/5).

.By this arrangement the delicate nerve matter of the brain and spinal cord lies between two layers of fluid—the internal layer of fluid being contained in the ventricles of the brain, and in the central canal of the spinal cord, and the external layer of fluid in the sub-arachnoid space. By means of these two 'water beds' the central nervous system is protected.

The functions of the cerebrospinal fluid. It acts as a buffer, protecting the brain and spinal cord. It conveys nourishment to the tissues of the central nervous system.

Lumbar puncture. Because the spinal cord ends at the level of the first or second lumbar vertebrae and the sub-arachnoid space extends to the level of the second sacral vertebra, a sample of cerebrospinal fluid may be drawn off by introducing a lumbar puncture needle into the *sub-arachnoid space* between these points—a process called lumbar puncture.

The examination of the cerebrospinal fluid thus obtained may reveal important information in conditions such as meningitis and sub-arachnoid cerebral haemorrhage.

THE DIFFERENT PARTS OF THE BRAIN

Development. The brain lies within the cranial cavity of the skull. It develops from a single tube which initially shows three enlargements, the fore-runners of the brain, termed, *fore-brain, mid-brain* and *hind-brain*. Thus:

The *Fore-Brain*, becomes the cerebral hemispheres, corpus striatum and the thalami

The *Mid-Brain*, the mid-brain ⎫ These three
The *Hind-Brain*, the pons Varolii, ⎬ form the
medulla oblongata, ⎭ *Brain Stem*
and the cerebellum.

(*see* page 338).

The **Cerebrum** fills the front and upper portion of the cranial cavity, termed respectively the anterior and middle cranial fossa (*see* Fig. 3/1, page 64). It consists of *two large hemispheres* of nerve cells (grey matter) and nerve fibres (white matter). The outer layer of grey matter is termed the *cortex* (*see* page 335). The two cerebral hemispheres are separated by a deep cleft, but united at their bases by the *corpus callosum*, a mass of white matter consisting of nerve fibres. Beneath this are islands of grey matter, the basal ganglia (*see* page 336).

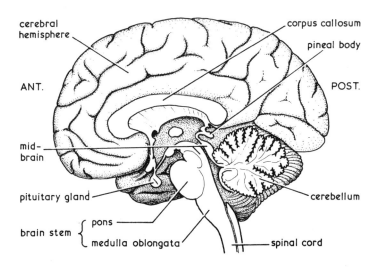

cerebral hemisphere

corpus callosum

pineal body

ANT.

POST.

mid-brain

pituitary gland

cerebellum

brain stem { pons

medulla oblongata

spinal cord

FIG. 22/6.—THE MEDIAL ASPECT OF THE RIGHT HALF OF THE BRAIN

Areas of the Brain. Fissures and sulci divide the cerebral hemisphere into areas. The cerebral cortex is arranged in convolutions or irregular folds in order to increase the expanse of grey matter. The depressions between the convolutions are called *sulci* and the deepest sulci form the *longitudinal* and *lateral* fissures. These fissures or sulci divide the brain into named areas or 'lobes' which correspond in position to the bones beneath which they lie, e.g. the frontal, temporal, parietal and occipital lobes.

The *longitudinal fissure* is a deep cleft in the medial plane separating the cerebrum into right and left cerebral hemispheres; into it dips a thin plate of dura mater called the *falx cerebri*. Similarly a thin partition of dura mater, the *falx cerebelli*, divides the cerebellum into right and left hemispheres.

The *lateral sulcus*, or the *fissure of Sylvius*, separates the temporal lobe from the frontal lobe (anteriorly) and from the parietal lobe more posteriorly (*see* Fig. 22/7, below).

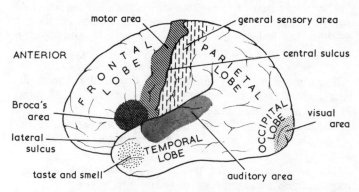

F<small>IG</small>. 22/7.—T<small>HE</small> L<small>ATERAL</small> A<small>SPECT</small> <small>OF THE</small> L<small>EFT</small> C<small>EREBRAL</small> H<small>EMISPHERE</small>. T<small>HE</small> C<small>ENTRAL</small> S<small>ULCUS</small> <small>OR</small> F<small>ISSURE</small> <small>OF</small> R<small>OLANDO</small> <small>SEPARATES THE</small> S<small>ENSORY</small> <small>AND</small> M<small>OTOR</small> A<small>REAS</small>

The *central sulcus* or fissure of Rolando separates the frontal from the parietal lobes. The occipital lobes of the cerebrum are situated behind the parietal lobes and rest upon the tentorium cerebelli—a fold of dura mater which separates the middle cranial fossa from the posterior cranial fossa below.

The **Cerebral Cortex** is composed of many layers of nerve cells; it is the *grey matter of the cerebrum*. It is arranged in irregular folds or convolutions, an arrangement which increases the surface area of the cerebral cortex, as scalloping a piece of material increases the length of its exact edge.

The *white matter* lies more deeply and consists of the nerve fibres belonging to the cells of the cortex.

The *cerebral cortex* is divided into various 'areas', some motor and some sensory in function.

The *motor area* lies just in front of the central sulcus (*see* Fig. 22/7), extending down as far as the lateral sulcus. This area of the cortex contains large cells which form the beginning of the *motor pathway* which controls movement of the opposite side of the body. The body is represented upside down—the lower limb, trunk, upper limb, neck and finally head controlling areas lie, from above down, in the motor area as indicated in Fig. 22/8.

The lowest part of the motor cortex is called *Broca's area* and is concerned with speech. Broca's area lies in the left cerebral hemisphere in right-handed people, and on the opposite side in those who are left-handed.

The *sensory cortex* lies immediately behind the central sulcus. Here the various modalities of sensation (*see* Figs. 22/7 and 22/8) are appreciated and interpreted.

The *auditory area* lies in the temporal lobe just below the longitudinal fissure. Here sound impressions are received and interpreted.

The *visual area* lies at the tip of the occipital lobe and receives images and impressions for interpretation.

The centres of *taste and smell* lie well forward in the temporal lobe.

The *white matter* of the cerebral hemispheres consists of nerve fibres running to and from the cortex linking up the various 'centres' of the brain with the spinal cord.

Basal Ganglia. As already mentioned, embedded in the mass of white matter of each cerebral hemisphere are certain small areas of grey matter, termed the *basal ganglia* or *nuclei*. Two of these are the *caudate* and *lentiform nuclei* and together form the *corpus striatum*. These structures are closely related to another mass of grey matter, the *thalamus*, which lies medially to them. It is likely that this system of nuclei and fibres which are part of the *extrapyramidal system* in some way influences tone and posture, integrates and co-ordinates the main voluntary muscle movements which are the concern of the great descending motor pathway, or the *pyramidal system*.

The *Thalamus* is chiefly concerned with the reception of sensory impulses, which may be either interpreted at a subcortical level, or relayed on to the sensory area of the cerebral

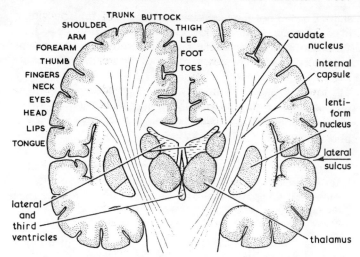

FIG. 22/8.—CORONAL SECTION OF BRAIN. MOTOR AREAS LABELLED ON THE LEFT, BASAL NUCLEI ON THE RIGHT

cortex. It appears to have an important regulating action on many of the highest centres for sensation and movement.

The *Hypothalamus*. In the region of the floor of the third ventricle are certain nuclei which have definite physiological activity. Some of them are related to the autonomic nervous system forming the 'highest part of that system'. Some nuclei also have connexions with the posterior lobe of the pituitary gland of the endocrine system on which they exert control. Functions such as body temperature regulation, hunger and thirst are regulated by centres in the hypothalamus.

Disorder in these areas leads to tremor at rest and rigidity of movement. **Parkinson's disease** or **paralysis agitans**, a progressive condition beginning in middle life, is a classical example. The head is flexed and held stiffly, the body bent, arms at the sides with fingers flexed, the thumb approximating the fingers in rhythmical pill-rolling movements. The thighs are slightly flexed and adducted; the patient takes little mincing steps. The skin of the face is smooth and unwrinkled giving a mask-like expression. Speech is slow and monotonous.

Surgery directed to the basal ganglia produces improvement in selected cases.

The Internal capsule is formed by fibres of the great motor and sensory pathways which link the cerebral cortex with the brain stem and spinal cord. In this part of their course these nerve fibres are closely packed together as they pass between the islands of grey matter.

Thrombosis of the artery supplying the internal capsule may lead to damage of the opposite side of the body (hemiplegia); such a cerebrovascular catastrophe constitutes a 'stroke'. (*See* Clinical Note, page 353.)

The Functions of the Cerebrum. These have been mentioned as the various parts have been described. To summarize: The cerebral cortex contains the higher centres controlling mental behaviour, thought, consciousness, moral sense, will, intellect, speech, language, and the special senses.

The cortex is the origin of all voluntary motor impulses controlling the skeletal muscles.

It is the final area for the reception of all incoming sensory nerve impulses and for their appreciation and interpretation, including skin sensation, touch, pain, pressure, temperature, vibration, texture, shape and size, and muscle and joint sense.

The Brain Stem is composed of the mid-brain, pons Varolii and medulla oblongata.

The *mid-brain* forms the upper part of the brain stem. Through it runs the *cerebral aqueduct* connecting the third and fourth ventricles (*see* Fig. 22/5, page 332). The mid-brain may be considered in two levels:

(1) The roof contains important reflex centres for sight and hearing.
(2) Through the base of the mid-brain the *great motor pathway* descends from the internal capsule above, to con-

tinue below through the pons and medulla on its way
to the spinal cord.

The *ascending sensory pathways*, on their journey from the
spinal cord, medulla and pons, pass through this portion of the

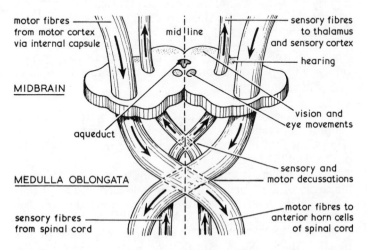

FIG. 22/9.—A SECTION OF THE MID-BRAIN SHOWING MAIN TRACTS OF FIBRES

mid-brain before entering the thalamus or internal capsule,
to reach their final distribution in the sensory cortex of the
cerebral hemispheres.

The mid-brain contains centres for the control of balance
and the movements of the eyes.

The *pons Varolii* forms the *middle portion of the brain stem*
and thus contains the same ascending and descending pathways
as the mid-brain. There are also many fibres running transversely through the pons which link the two lobes of the cerebellum; and the cerebellum with the cerebral cortex.

The *medulla oblongata* forms the *lower portion of the brain stem* linking the pons with the spinal cord. The medulla lies in the posterior cranial fossa and joins the spinal cord just below the foramen magnum of the occipital bone.

The main features of the medulla are that here the descending motor pathways cross from one side of the brain stem to the other. This is called the *motor decussation*. A similar arrangement of the sensory pathways occurs in the medulla and is referred to as the *sensory decussation*.

The medulla contains the nuclei or cell bodies of several important cranial nerves. It also contains certain 'vital centres' which control respiration and the cardiovascular system. Injury to this part of the brain stem is therefore liable to have very serious consequences.

The **Cerebellum** is the largest part of the hind-brain. It occupies the posterior cranial fossa and is roofed over by the *tentorium cerebelli*, a fold of dura mater which separates it from the occipital lobes of the cerebrum.

The cerebellum is separated from the pons and medulla by the cavity of the fourth ventricle. It is divided into two hemispheres, right and left, by a deep cleft into which dips another fold of dura mater, the *falx cerebelli*.

The arrangement of grey and white matter is similar to that found in the cerebrum with the grey matter arranged at the surface. The surface is ridged rather than folded into convolutions, the fissures between the ridges being very much closer together than the sulci of the cerebral cortex.

The cerebellum has connexions with many other parts of the nervous system. Its principal connexions are with the cerebral hemisphere of the opposite side and with the brain stem. It also receives fibres from the spinal cord and is connected with the reflex centres of sight in the roof of the midbrain, with the thalamus and with the auditory or acoustic nerve of hearing.

The **function of the cerebellum** is to regulate posture and postural activities. It plays an important part in muscular coordination and the maintenance of balance. Whereas the cortico-spinal fibres running between the cerebral cortex and the spinal cord cross (*see* above), and thus the cerebral cortex controls the movements of the opposite side of the body, the cerebellar hemisphere controls muscle tone and posture on its own side.

A unilateral lesion of the cerebellum causes disturbance of posture and muscle tone. Movement is very inco-ordinate, a patient may be unable to put food into his mouth and smears it across his face; he sways in walking and tends to fall towards the affected side. All voluntary movement is slow, and the muscles of the limbs are limp and flail-like. Speech is slow.

THE CRANIAL NERVES

There are twelve pairs of cranial nerves. Some are mixed nerves, i.e. both motor and sensory, some motor only, and some sensory nerves, e.g. the nerves of the special senses.

(1) *Olfactory* (sensory), the nerve of smell. (*See also* pages 369–70.)

(2) *Optic* (sensory), the nerve of sight. (*See also* pages 371–77.)

(3) *Oculo-motor* nerve supplies most of the external muscles of the eye. It also transports parasympathetic nerves to supply the ciliary muscle and the muscles of the iris. **Clinically** complete division of this nerve results in *ptosis*, *squint*, and *loss of the reflexes to light and accommodation.*

(4) *Trochlear* (motor), to one muscle of the eye, the external oblique.

(5) **Trigeminal.** This is the largest cranial nerve. It is principally a *sensory nerve*, supplying most of the skin of the head and face; the membranes of the mouth, nose and paranasal sinuses and the teeth and, by means of a small *motor branch*, the muscles of mastication. It is divided into three main branches, passing forwards from the trigeminal ganglion: the ophthalmic, maxillary and mandibular nerves, to supply sensation to the respective areas of the face, mouth, teeth and part of the scalp as indicated in Fig. 22/10. It is also concerned with supplying taste sensation to the tongue.

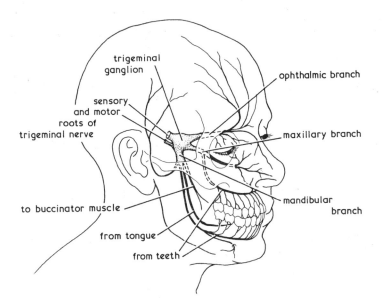

FIG. 22/10.—DISTRIBUTION OF THE TRIGEMINAL (5TH CRANIAL) NERVE

Clinical Notes

Herpes, involving the ophthalmic branch, is serious when it invades the cornea and results in scarring, as this may cause partial or complete blindness. Any herpetic lesion may result in *post-herpetic neuralgia* characterized by persistent severe pain. Much more rarely post-herpetic encephalitis may occur.

Trigeminal neuralgia may affect the distribution of one or all of the three branches. The pain is acute and paroxysmal, and in some cases very prostrating. The condition can be relieved by certain surgical measures, but this numbs the face and causes discomfort and is not lightly undertaken.

(6) *Abducens* (motor), to one muscle of the eyeball, the lateral rectus.

(7) *Facial*. This nerve is mainly *motor* to the muscles of expression of the face, and to the scalp (*see* Fig. 22/11). It is a *sensory* nerve in that it is concerned with conveying taste sensation from the tongue (*see* page 368).

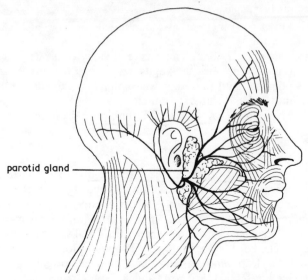

parotid gland

FIG. 22/11.—DISTRIBUTION OF THE FACIAL (7th CRANIAL) NERVE TO THE MUSCLES OF THE FACE

Clinical aspects. Paralysis of the facial nerve may be due to a number of causes, including fractures of the skull, tumours and poliomyelitis affecting the brain stem.

Bell's palsy is an acute lower motor nerve lesion. The affected side of the face is motionless, the eye stays open, tears flow over the face, food accumulates in the cheek. Little is known of the cause of Bell's palsy. Most cases recover completely.

(8) *Auditory* or *Acoustic* (sensory), the nerve of hearing. This nerve is in two parts, the *cochlear nerve*, the true nerve of hearing, and the *vestibular nerve*, which is concerned with equilibrium. (*See also* page 385.)

(9) *Glosso-pharyngeal* contains motor and sensory fibres. It is *motor* to one of the constrictor muscles of the pharynx, *secreto-motor* to the parotid gland and *sensory* to the posterior third of the tongue and part of the soft palate.

(10) The *Vagus* nerve is composed of motor and sensory fibres; its functions are mentioned on page 362. (*See also* Fig. 23/2, page 365.)

(11) *Spinal Accessory.* This nerve divides into two parts; one accompanies the vagus, passing to the larynx and pharynx, and the other is a *motor nerve* to the sterno-mastoid and trapezius muscles.

(12) *Hypoglossal* (motor), to the muscles of the tongue.

THE SPINAL CORD

The **spinal cord** begins at the medulla oblongata where it emerges from the foramen magnum and ends between the first and second lumbar vertebrae, where it tapers as the *conus medullaris* and from this a thin prolongation of the pia mater, the *filum terminale* which has pierced the dural sac, runs to the coccyx. The cord is about 45 cm (18 inches) long, divided in front by a deep *anterior fissure* and by a narrow one at the back (*see* Fig. 22/12). See also Clinical Notes, page 357.

The cord has two enlargements, cervical and lumbar, from which emerge the nerve plexuses to supply the upper and lower limb; those from the thoracic region form the intercostal nerves.

A transverse section of the cord (*see* Fig. 22/12) shows the arrangements of grey matter in the form of a letter H. The

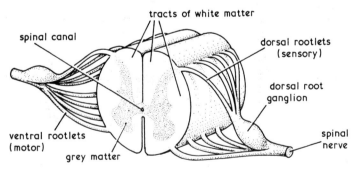

FIG. 22/12.—DIAGRAM OF A SECTION OF THE SPINAL CORD

spinal canal, with its cerebrospinal fluid, passes through the centre.

The *cauda equina* is so called from its resemblance to a tail—*cauda* and *equina*—of a horse (*see* Fig. 22/13); it is the sheath of roots of the spinal nerves passing down the spinal canal from their attachment to the spinal cord to their point of emergence through the intervertebral foramina.

The functions of the spinal cord are (*a*) communication between the brain and all parts of the body, and (*b*) reflex action.

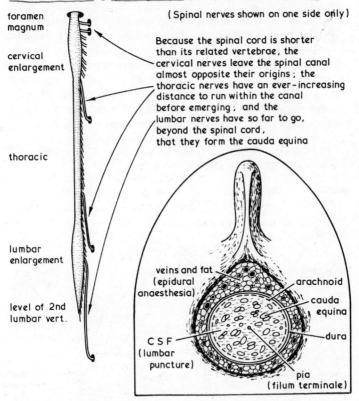

(Spinal nerves shown on one side only)

Because the spinal cord is shorter than its related vertebrae, the cervical nerves leave the spinal canal almost opposite their origins ; the thoracic nerves have an ever-increasing distance to run within the canal before emerging ; and the lumbar nerves have so far to go, beyond the spinal cord, that they form the cauda equina

foramen magnum

cervical enlargement

thoracic

lumbar enlargement

level of 2nd lumbar vert.

veins and fat (epidural anaesthesia)

arachnoid

cauda equina

C S F (lumbar puncture)

dura

pia (filum terminale)

FIG. 22/13.—THE SPINAL CORD. THE INSET SHOWS A TRANSVERSE SECTION OF THE SPINAL CANAL IN THE FOURTH LUMBAR VERTEBRA

Spinal Nerves. The thirty-one pairs of spinal nerves arise segmentally by two roots, anterior and posterior. *Motor nerve fibres forming an anterior root* unite with the *sensory nerve fibres of a posterior root* to form a *mixed spinal nerve* (*see* Fig. 22/12). This union takes place before the nerve passes through the intervertebral foramen, but soon after emerging it divides again into anterior and posterior primary divisions.

The *posterior primary divisions* supply the skin and muscles of the back; the *anterior primary divisions* form branches which become the nerve plexuses for the limbs (*see* page 352) and, in the thoracic region, the intercostal nerves, already mentioned.

See Clinical Note on lesions of these peripheral nerves, page 358. (*See also* motor and sensory pathways, pp. 338–9, 348–50).

A Reflex Action requires the following structures which form a reflex arc:

A sensory organ which receives the impulse, e.g. the skin.

A sensory nerve fibre which conducts this impulse to the cells in the posterior root ganglion and thence by their fibres to the grey matter of the posterior horn of the spinal cord.

The spinal cord where connector nerves pass impulses on to the anterior horn of the cord.

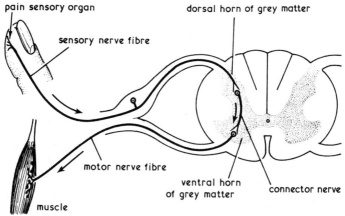

FIG. 22/14.—DIAGRAM OF THE PARTS REQUIRED IN THE FORMATION OF A REFLEX ARC

A motor nerve cell in the anterior horn of the spinal cord which receives and transmits the impulse along motor nerve fibres.

A motor organ, e.g. a muscle, which, stimulated by the motor nerve impulse, performs the action.

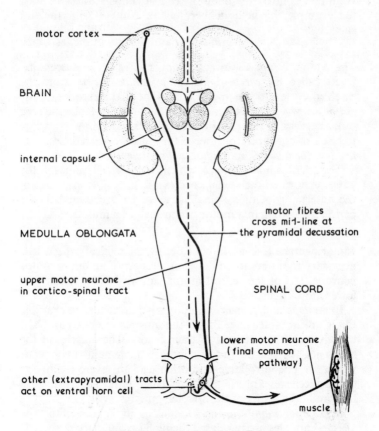

FIG. 22/15.—THE MOTOR NERVE TRACT

Upper motor nerves commence in the cerebral cortex and pass into the spinal cord, crossing at the medulla. As indicated, these nerves converge to pass, closely packed together with the nerves of the ascending sensory tract, through the internal capsule (*see* page 338) (*see also* Clinical Note, page 353, on the result of a cerebrovascular accident in this area).

Lower motor nerves begin as cells in the grey matter of the cord, and pass out from the anterior horns to supply the muscles and other structures.

Reflex actions are part of the defence mechanism of the body, and take place much more rapidly than voluntary actions, e.g. the closing of the eye when irritated by dust, the movement of withdrawing the hand from some article accidentally touched if unpleasantly hot. Reflex actions can be inhibited by voluntary control; the hand instead of being withdrawn may, for example, be held deliberately in contact with the hot surface.

The Motor Nerve Pathways. Impulses travel in descending tracts called the *cerebrospinal* or *pyramidal tracts*, from the cerebral cortex to the spinal cord. The first neurones, *upper motor neurones*, have their cell bodies in the pre-Rolandic area of the cerebral cortex (*see* Fig. 22/15) and many fibres converge to be closely grouped together as they pass between the *caudate and lentiform nuclei*, in the *internal capsule*.

The *lower motor neurones*, which begin as cell bodies in the anterior horn of the spinal cord (*see* Fig. 22/14), pass out in the anterior root of a spinal nerve to be distributed to the periphery, ending in a motor organ such as a muscle.

Motor neurone lesions. In considering the **clinical aspect**, it is necessary to differentiate between a lesion of an upper motor neurone, i.e., of the central motor pathway and a lesion of a lower motor neurone.

In an *upper motor neurone lesion*—hemiplegia is an example (*see* Clinical Note, page 357)—the muscles are not paralysed but are weak and control of them is lost. The muscles of the limbs may be spastic and involuntary movements may occur which are uncontrollable and often lead to severe rigidity in spasm. Reflexes are exaggerated. There is no loss of muscle tone and no wasting of the affected muscles.

In a *lower motor neurone lesion*, as in poliomyelitis, the affected muscles are paralysed, being limp and flaccid, there is wasting and normal reflexes are lost; if the subject is a child the limb may not develop.

Bell's palsy is another example (*see* Clinical Note on page 343).

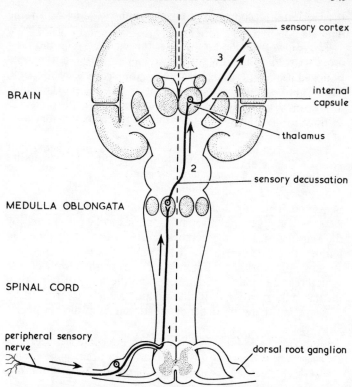

FIG. 22/16.—THE SENSORY NERVE TRACT HAS THREE RELAYS

1. From the periphery to the spinal cord, the axons ascending as far as the nuclei in the medulla.
2. From the grey matter of the medulla to the thalamus.
3. From thence to the post-Rolandic sensory area (*see* Fig. 22/7).

Sensory Nerve Pathway. The sensory nerve impulses travel in ascending tracts which consist of a three-neurone pathway.

The *first or most peripheral neurone* has its cell body in the sensory ganglion of the posterior nerve root of a spinal nerve; one branch, a *dendron*, passes to the periphery to end in some *sensory* organ *such as the skin* (*see* note, p. 350); the other branch, the *axon*, passes into the spinal cord and ascends in the posterior column to arborize round a nucleus in the medulla.

The second neurone cell arises in the same nucleus and then crosses the mid-line in a similar way as the descending motor pathway (*see* Fig. 22/16), to form the *sensory decussation*, ascends through the pons and mid-brain to reach the thalamus.

The *third and final neurone* commencing in the thalamus passes through the internal capsule to reach the sensory area of the cerebral cortex.

These ascending tracts convey impulses of touch, joint position and vibration sensation; others convey impulses of touch, pain and temperature.

Sensation. Thus a *peripheral sensory nerve* as depicted in Fig. 22/17 will carry some 'afferent' impulses to be interpreted by the sensory area in the cerebral cortex as touch, pain, itch, temperature and warm and cold sensations from superficial structures, and other 'afferent' impulses arising from deeper structures as in pain, pressure etc. and the sense of the movement and the position of joints and muscles. The interpretation of sensation therefore depends on stimuli from the periphery,

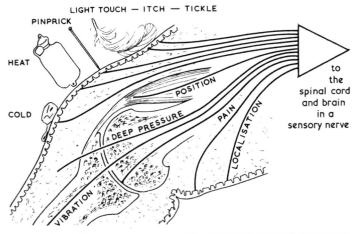

FIG. 22/17.—A DIAGRAM SHOWING THE VARIETIES OF SENSATIONS COLLECTED FROM SUPERFICIAL AND DEEP STRUCTURES

relayed by several neurones reaching eventually the central interpreting station in the brain.

Nerve synapse. It has been noted (*see* page 327) that the *axon* of a nerve is the emitting fibre and the dendrites (there are more than one) the fibres which receive the nerve impulses and pass them on to the *nerve cell*. In the central nervous system impulses may be passed along relays of neurones, as instanced on the ascending sensory neurones, Fig. 22/16. It is considered that the process of passing an impulse on is effected without actual continuity of structure. This process is indicated in the accompanying diagram where the *synaptic junction*, as it is called, is shown.

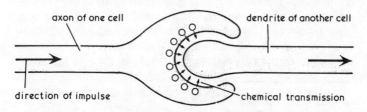

axon of one cell

dendrite of another cell

direction of impulse

chemical transmission

FIG. 22/18.—NERVE SYNAPSE

THE MAIN NERVE PLEXUSES
AND THEIR TRUNKS

The anterior primary divisions of the spinal nerves, other than those which arise in the thoracic region and form the intercostal nerves, are arranged into four main plexuses.

The Cervical Plexus is formed by the first four cervical nerves. It lies in the neck beneath the sternomastoid muscle. Many branches arise from it to supply some of the muscles of the neck. The *phrenic nerves* which supply the diaphragm arise from this plexus.

The Brachial Plexus is formed by the *four lower cervical nerves and the first thoracic nerve*. It is situated in the posterior triangle of the neck behind the clavicle and in the axilla. At first, *three trunks* are formed; these then divide and unite again to form *three cords*, *lateral*, *medial*, and *posterior*. From these cords 5 principal nerves arise which supply the arm and some of the neck and chest muscles (*see* Fig. below).

The **Lumbo-Sacral Plexus** (*see* Fig. 22/20) provides the principal spinal nerves to the lower limb.

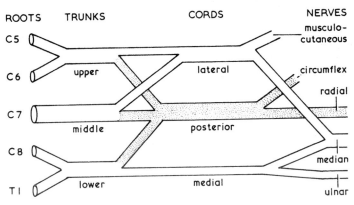

FIG. 22/19.—THE BRACHIAL PLEXUS (LEFT) SHOWING THE ORIGIN OF THE PRINCIPAL NERVES OF THE UPPER LIMB

The posterior divisions of the trunks are shaded.

The **Lumbar Plexus** from the first 4 lumbar nerve roots lies in the psoas muscle supplying it and divides into two branches, the *femoral nerve* passing beneath the inguinal ligament, through the femoral triangle (*see* page 145) to supply the muscles on the front of the thigh, and the *obturator nerve* which enters the thigh through the obturator foramen to supply the muscles on the inner side of it.

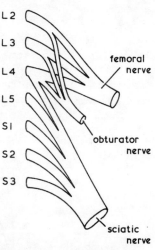

The **Sacral plexus** consists of the 4th and 5th lumbar nerves and the sacral nerves uniting to form the great *sciatic nerve* which passes into the thigh through the great sacral notch supplying the hamstring muscles. It then divides into the *medial* and *lateral popliteal nerves* (*see* Fig. 22/23) which supply the muscles on the back of the thigh and all the muscles, back and front, below the knee.

FIG. 22/20.—THE LUMBO-SACRAL PLEXUS, SHOWING THE ORIGIN OF THE PRINCIPAL NERVES TO THE LOWER LIMB

Clinical Notes

Cranial nerves. Disease or damage to the cranial nerves causes the following symptoms:

i.	Loss of smell
ii.	Dimness or loss of vision
iii, iv, vi.	Double vision, squint
v.	Pain or loss of feeling on the face, toothache, and weakness of mastication (*see* Clinical Note, page 342)
vii.	Paralysis of the facial muscles (*see* Clinical Note, page 343)
viii.	Auditory deafness or tinnitus, vestibular giddiness, loss of balance
ix, x, xi.	Difficulty in swallowing
xii.	Weakness of the tongue, causing difficulty in chewing and speaking

Cerebral hemispheres.

Cerebral lesions. The disease or damage which presents after injury, or following a *cerebrovascular accident* to the brain, depends on the areas

(*contd. on p. 357*)

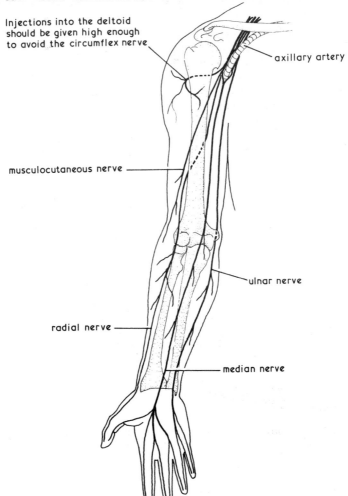

Injections into the deltoid
should be given high enough
to avoid the circumflex nerve

axillary artery

musculocutaneous nerve

ulnar nerve

radial nerve

median nerve

FIG. 22/21.—SHOWING THE MAIN NERVES ARISING FROM THE BRACHIAL
PLEXUS

The *radial nerve* may be damaged in fractures of the humerus, when the branch supplying
the extensors of wrist and fingers may be injured or for any other cause put out of action.
The result is *wrist drop* which requires splintage and physiotherapy.

Injury to the *median nerve* will destroy sensation to the skin of the first $3\frac{1}{2}$ digits, and to
the *ulnar nerve* half the ring finger and the little finger ($1\frac{1}{2}$ digits). Wasting and loss of power
in the muscles of the thenar and hypothenar eminences and in the small muscles of the palm
result from injury to these two nerves.

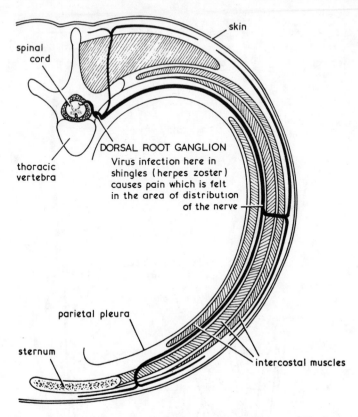

FIG. 22/22.—DIAGRAM OF THE COURSE OF THE SENSORY PART OF A THORACIC NERVE

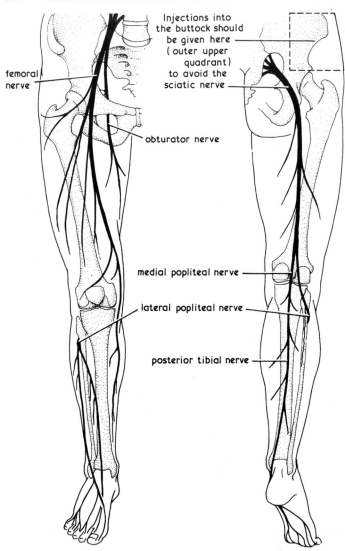

femoral nerve

Injections into the buttock should be given here (outer upper quadrant) to avoid the sciatic nerve

obturator nerve

medial popliteal nerve

lateral popliteal nerve

posterior tibial nerve

FIG. 22/23.—THE PRINCIPAL NERVES FROM THE LUMBO-SACRAL PLEXUS (*see* FIG. 22/20) INDICATING THEIR COURSE THROUGH THE LOWER LIMB

(*contd. from p. 353*)
and neurones involved: motor and sensory nerves pass through the internal capsule (*see* pages 338, 347) on their journey to and from the brain, so both may be affected.

Motor paralysis of the spastic type with muscular rigidity and increased reflexes follows involvement of the upper motor neurones (*see* page 348). This may affect only the arm and leg of one side—*hemiplegia*—but muscles of the face, head, neck and trunk, though the latter often escapes, may also be involved.

Sensory paralysis follows injury to the sensory path. *Reflex actions* are abnormal. These involve the *organic reflexes* of the pupil of the eye, which may be contracted, or fail to contract; the reflexes of the bladder are affected causing paralysis of the sphincters and the bladder wall which results in retention of urine with overflow; involvement of the rectum, with disorder of the reflex of defaecation, may also occur and these matters are of great practical importance in nursing care.

As the cerebral hemispheres are also the part of the brain in which highly developed functions such as speech, vision, taste and smell and memory are situated, damage can cause many symptoms.

Cerebral lesions need the services of an experienced physiotherapist in order to re-educate and train the parts which remain and initiate other tracts to function. It also needs the continuous interest and co-operation of doctors and nurses, and most important too the full collaboration of the patient and his relatives in order to ensure the highest degree of rehabilitation possible for him.

Basal ganglia. Parkinson's disease, paralysis agitans or shaking palsy, is thought to be due to degeneration of the basal ganglia (*see* Clinical Note, page 337).

Lesions in one of the **cerebellar hemispheres** give rise to symptoms on the same side of the body (*see* Clinical Note, page 341).

Brain stem, pons and medulla. The vital centres controlling breathing and blood pressure are located here, so severe damage causes death. The number of nervous pathways concentrated in this region is so great that even small lesions cause much weakness and loss of feeling.

Spinal cord lesions

Division or transection of the spinal cord due, most often, to a traffic accident, is a serious injury which may be complete or partial. A complete transection is considered here. The higher the injury, the greater is the disablement. In the *cervical region* arms, trunk and legs are affected (the patient is helpless). When the phrenic nerve escapes, the diaphragm may be unaffected; if involved artificial respiration is needed until a mechanical respirator can be employed.

A *transection in the dorsal and lumbar regions* results (in the dorsal) in paralysis of some of the intercostal muscles, the abdominal muscles and in those of both the lower limbs with involvement of the sphincters of urethra and rectum (*see* page 358).

Spinal shock occurs at first, lasting for about a week. The muscles of the areas below the injury are *limp and flaccid*, as in a lower motor neurone lesion (*see* page 348); all *reflexes are abolished*, there is *incontinence of urine and faeces*. During this stage there is *complete anaesthesia* of all areas below the lesion, with danger of injury to the skin, so that expert routine nursing care is essential.

Spasticity and rigidity. The state of flaccid paralysis passes, muscles regain tone but are weak, the affected limbs become rigid and spastic, reflex movements, particularly affecting flexor and adductor muscle groups, occur, but there is no voluntary control over these movements. This power is lost. At this stage deformities are likely to occur.

The condition of the bladder and rectal function needs watching. Complete urinary incontinence gives place to *retention of urine with overflow*, when either regular catheterization or continuous bladder drainage will be employed as *residual urine in the bladder* is a source of infection which may spread to the entire urinary tract.

With a co-operative patient it may be possible to establish automatic reflex bladder and bowel action.

Similar symptoms may arise in any condition causing interruption of the spinal cord such as pressure by a tumour, or in neurological disease such as multiple sclerosis.

A patient facing the grave disability brought about by a *transection of the spinal cord* needs all the understanding and help we can give, for his collaboration is essential. It is ideal if he can be nursed in a special 'spinal' unit, such as that at Stoke Mandeville, where his needs will be understood, all facilities provided for moving, turning, lifting, etc., and his treatment complemented by physiotherapy, and recreational, educational and occupational measures geared to the best possible degree of rehabilitation for him; a remarkable amount can be achieved.

Peripheral nerve lesions may be due to pressure on a nerve root or roots, causing inflammation (radiculitis). Some disorder of the spine, an intervertebral disc lesion, spondylosis, tumour and spinal fracture may be the cause. The presence of a cervical rib, causing *brachial neuritis* is mentioned on page 76. An intervertebral disc lesion is a common cause of *sciatic neuritis* (sciatica).

Division of any of these **mixed nerves,** which may occur in road accidents, will deprive the areas supplied by them of the power of movement as this constitutes a *lower motor neurone lesion* (*see also* page 348) and of sensation. *Peripheral nerve injuries* can be surgically repaired, but it takes a long time for one of the principal nerves of a limb to grow and regenerate; in the meantime, physiotherapy is employed to assist the process and keep the affected muscles in tone.

Neuritis is a composite term used to indicate disorder of a peripheral nerve from any cause, whether inflammation, some form of poisoning, as in alcoholic neuritis, or pressure. *Symptoms of the inflammatory type* are variable; there is usually pain which is worse at night and not relieved by rest. Disturbances of sensation include numbness and tingling; in some instances paralysis occurs. In *polyneuritis the condition is symmetrical— examples include alcoholic neuritis, diabetic neuritis and neuritis due to metabolic disturbances including dietetic and vitamin deficiencies, for example in beri-beri.*

When due to pressure there is generally paresis or paralysis, but pain may not be constant. This type is named after the plexus or nerve involved as:

Brachial plexus neuritis may be due to infection, injury or pressure.

Radial nerve neuritis (*see* note on page 354, Fig. 22/21). The radial nerve may be injured if the arm is allowed to hang over the side of a stretcher or operating table.

Ulnar nerve pressure may arise from leaning on the elbow in lying.

Median nerve compression in the carpal tunnel is mentioned on page 97.

Sciatic neuritis—*sciatica* is thought in many cases to be due to pressure from a prolapsed intervertebral disc or other lesion of the lower part of the vertebral column.

The *lateral popliteal nerve* may be compressed when the leg is in plaster of paris as it winds round the head of the fibula (*see* Fig. 22/23, page 356).

Encephalitis is inflammation of the substance of the brain and is usually due to viral infection.

Meningitis is inflammation of the meninges of the brain (*see* Clinical Note on page 331).

Neuro-surgery is a highly specialized branch which includes all operations undertaken on the brain, spinal cord, and the peripheral nerves. Careful pre-operative investigations are carried out in order to determine the exact site of the lesion or tumour, to assess as far as possible the prognosis, and to decide what after-care and follow-up will be necessary.

Craniotomy is opening the skull, generally to deal with a tumour, blood or blood-clot, or a depressed fracture of the vault, causing pressure on the brain. This subject is too comprehensive to be dealt with here; but the reader is referred to the clinical notes on *head injuries* on page 73 and to those on spinal cord and peripheral nerve lesions dealt with above.

Chapter 23

THE AUTONOMIC NERVOUS SYSTEM

The *autonomic nervous system* is dependent on the central nervous system with which it is connected by afferent and efferent nerves. It behaves as if it were a part of the central nervous system which had migrated from it in order to reach glands, blood vessels, heart, lungs and intestine. Because the autonomic nervous system deals mainly with involuntary or automatic nervous control of viscera it is sometimes called the *involuntary nervous system*. The autonomic nervous system is divided functionally into two parts:

(a) the *sympathetic system* (*see* below) which lies in front of the vertebral column and is associated and connected with the spinal cord by nerve fibres.

(b) the *parasympathetic system* which is divided into two parts composed of the cranial and sacral autonomic nerves (*see* Fig. 23/2).

The Sympathetic System consists of a *double chain of ganglionated cords* extending from the base of the skull, lying in front of the vertebral column, to end in the pelvis opposite the coccyx as the *ganglion impar*. These ganglia are arranged in pairs and distributed from the following regions:

In the neck:	Three pairs of *cervical ganglia*.
In the chest:	Eleven pairs of *thoracic ganglia*.
In the loins:	Four pairs of *lumbar ganglia*.
In the pelvis:	Four pairs of *sacral ganglia*.
Front of coccyx:	The *ganglion impar*.

These ganglia are intimately connected with the central nervous system through the spinal cord by means of communicating branches, which pass outwards from cord to ganglia, and inwards from ganglia to cord.

Other sympathetic ganglia are placed in relation to these

two great chains of ganglia and with their fibres form the sympathetic plexuses (*see* Fig. 23/1).

(1) *The cardiac plexus* is placed near the base of the heart and sends branches to it and to the lungs.

(2) *The coeliac plexus* lies behind the stomach and supplies organs in the abdominal cavity.

(3) *The mesenteric plexus* lies in front of the sacrum and supplies organs in the pelvis.

Functions. Sympathetic nerves supply innervation to the muscle of the heart, the involuntary muscle of all blood vessels, and of viscera such as the stomach, pancreas and intestines. It supplies motor secretory fibres to the sweat glands, motor fibres to the involuntary muscle in the skin—the *arrectores pilorum*—and maintains the tone of all muscle, including the tone of voluntary muscle.

The Parasympathetic System. The *Cranial Autonomics* are the third, seventh, ninth and tenth cranial nerves. These form the means by which the parasympathetic fibres pass out from the brain to the organs partly controlled by them.

By means of the *third cranial nerve*, the *oculomotor nerve*, fibres reach the circular muscular fibres of the iris, stimulating the movements which determine the size of the pupil of the eye.

By the *seventh nerve*, the *facial*, and the *ninth*, the *glosso-pharyngeal*, motor secretory fibres reach the salivary glands.

The *vagus* or *tenth cranial* nerve is the largest autonomic nerve. It has a very wide distribution and sends fibres to a number of glands and organs as indicated in Fig. 23/2. This distribution is closely associated with that of the sympathetic fibres (*see* system of dual control of certain organs below).

The Sacral Parasympathetic Nerves pass out from the sacral region of the spinal cord. These form the nerves to the pelvic viscera and together with the sympathetic nerves form the plexuses which supply the colon, rectum and bladder.

A System of Dual Control (sympathetic and parasympathetic). Although some organs and glands have only one source of supply, sympathetic or parasympathetic, these are in the minority; the majority have a dual supply, receiving some fibres from the sympathetic system and some from the cranial or sacral autonomic nerves, the activity of the organ being stimulated by one set of nerves and retarded or inhibited by the other set—each acting in antagonism to the other. In this way an exact adjustment is maintained between activity and rest, and the smooth rhythmic action of the internal organs, glands, blood vessels, and involuntary muscle is maintained.

Thus the heart receives *accelerator fibres* from the *sympathetic nerves*, and *inhibitory fibres* from the *vagi*.

The blood vessels have their *vasoconstrictors* and *vasodilators*.

The alimentary canal has *accelerator* and *inhibitory* nerves, which increase and decrease peristaltic movements respectively (*see* plan below).

Organ	Action increased or Activated by:	Action depressed or Inhibited by:
Heart	Sympathetic (rate and force increased)	Vagus (rate and force decreased)
Bronchi	Vagus (constricted)	Sympathetic (dilated)
Stomach	Vagus (contracted)	Sympathetic (relaxed)
Intestine	Vagus (contracted)	Sympathetic (relaxed)
Bladder	Sacral autonomic (contracted)	Sympathetic (relaxed)
Pupil of eye (iris)	3rd cranial autonomic (contracted)	Sympathetic (dilated)

In the case of an organ which possesses a sphincter muscle, such as the stomach in the *pyloric sphincter*, the intestine in the *ileocolic sphincter*, and the bladder in the *internal urethral sphincter*, the nerve which causes contraction of the organ inhibits the sphincter and vice-versa. For example, in the act of micturition the urethral sphincter is relaxed whilst the muscle in the wall of the bladder is contracted thus enabling the bladder to be emptied.

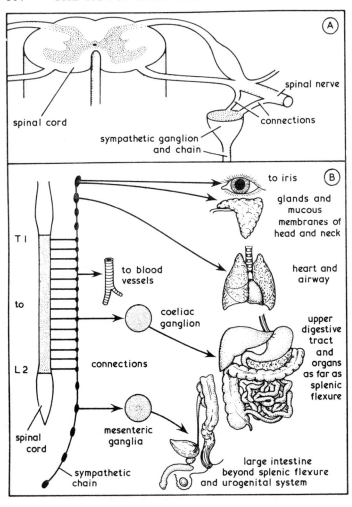

Fig. 23/1.—(A) The Connections between Spinal Cord, Spinal Nerve and Sympathetic Chain (B) Diagram of One Ganglionated Cord of the Sympathetic System, the Principal Plexuses and some of the Organs supplied by Them

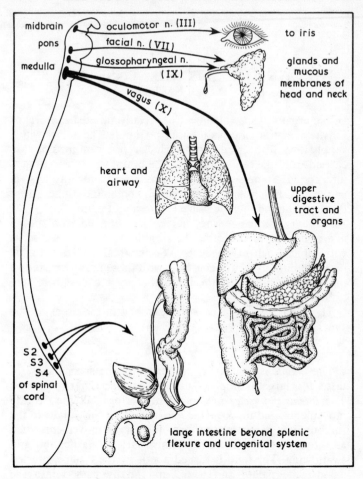

FIG. 23/2.—THE PARASYMPATHETIC SYSTEM. DIAGRAM OF THE ORGANS
SUPPLIED BY THE CRANIAL AND SACRAL AUTONOMIC NERVES

Chapter 24

ORGANS OF SPECIAL SENSE—TASTE AND SMELL

The organs of special sense are specially adapted end-organs for the reception of certain kinds of stimuli. The nerves which supply them form the means by which sensory impressions are carried from the sense organs to the brain, where sensation is interpreted. Some sense impressions arise from outside such as those of touch, taste, sight, smell, and sound. Others arise from within and include hunger, thirst, and pain.

In each case the sensory nerves are supplied with special nerve endings for collecting the stimuli of the particular sense with which each organ deals. We apparently taste with the nerve endings in the tongue, hear with those in the ear, and so on, but in reality it is the brain that appreciates these sensations.

The *sense of touch* has been described in the notes on the skin (page 290).

TASTE

The tongue is principally concerned in the special sense of taste. It is largely composed of muscle which is in two groups. The *intrinsic muscles of the tongue* perform all the delicate movements, and the *extrinsic muscles* attach the tongue to the surrounding parts and perform the larger movements such as those which form an important part of mastication and swallowing. The food is turned about by the tongue, pressed against the palate and teeth, and finally passed into the pharynx.

The tongue lies in the floor of the mouth, at its *root* the vessels and nerves pass in and out, the *tip* and *margins of the tongue* are in contact with the lower teeth, and the *dorsum* is the arched surface on top of the tongue. When the tongue is turned up, the under surface, the *frenulum linguae*, a soft ligamentous structure which attaches the posterior part of the tongue to the floor of the mouth, can be seen (Fig. 14/2, page 213). The anterior portion of the tongue is free. When protruded the tip of the tongue becomes pointed, but when lying in the floor of the mouth and relaxed the tip is rounded.

The *mucous membrane of the tongue* is moist and pink in health. On the upper surface it has a velvety appearance and is covered by papillae, of which there are three varieties.

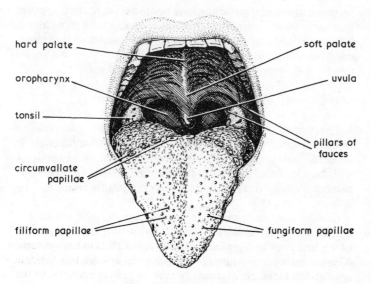

FIG. 24/1.—THE MOUTH AND TONGUE

Circumvallate papillae. Of these there are from eight to twelve placed at the base of the tongue. These are the largest and each one is surrounded by a little moat-like depression. These papillae are arranged in a V-shape at the back of the tongue.

Fungiform papillae are distributed over the tip and sides of the tongue, and are fungoid in shape.

Filiform papillae are the most abundant and are found over the whole surface of the tongue.

The end-organs of taste are the **taste buds,** which are very numerous in the walls of the circumvallate and fungiform papillae. The filiform papillae are concerned more with the sense of touch rather than actual taste. Taste buds are also contained in the mucous membrane of the palate and pharynx.

There are four true **sensations of taste:** sweet, bitter, sour, and salt. Most foods have **flavour** or aroma as well as taste, but this stimulates the nerve endings of smell and not those of taste. All food must be in liquid form before it can be tasted, and it must actually come into contact with the nerve endings capable of receiving the different stimuli. Different taste buds give rise to different tastes.

The tongue has a complicated **nerve supply.** The muscles of the tongue are innervated by the *hypoglossal* (twelfth cranial) nerve. *Sensation* is divided into 'general sensation', which deals with tactile sense, discrimination of size, shape, texture, consistence, temperature, etc., and 'special taste sensation' (*see* above).

General sensation impulses from the anterior part of the tongue travel in the *lingual nerve*, a branch of the *fifth cranial nerve*, and *special taste* impulses travel in the *chorda tympani* which runs with the lingual nerve but later joins the *seventh cranial*, the facial nerve.

The *glossopharyngeal* nerve, ninth cranial, carries both general sensation and special taste sensation impulses from the posterior one-third of the tongue.

Thus the sensations of taste of the tongue are supplied by the fifth, seventh and ninth cranial nerves, and its movements are innervated by the twelfth cranial nerve.

Clinically the sense of taste, like that of smell (*see* opposite) is delicate and can be diminished by a cold in the head or disorder of the mouth, stomach and intestinal tract. A physician looks at the tongue and can be helped by a nurse who examines it carefully, observing whether it is dry or moist, large, flabby and pale, or small and red, furred, cracked or fissured.

Glossitis, or inflammation of the tongue, may be acute, with the tongue covered by ulcers and tender; or chronic. This form is seen in patients with chronic indigestion or infected teeth. The tongue is flabby and pale, with the edges indented by pressure from the teeth. As a rule, with improvement in the general health and care in the hygiene of the mouth, chronic glossitis clears up.

Leukoplakia is marked by thick white patches on the tongue (and also on the buccal mucous membrane and the gums). It is seen in smokers.

SMELL

The olfactory or first cranial nerve supplies the end-organs of smell. The filaments of this nerve arise in the upper part of the mucous membrane of the nasal cavities which is known as the *olfactory portion of the nose*. It is lined with highly special-ized cells from which minute fibrils pass to arborize with fibres from the olfactory bulb. The *olfactory bulb*, which is actually

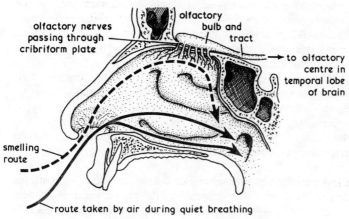

olfactory nerves passing through cribriform plate

olfactory bulb and tract

to olfactory centre in temporal lobe of brain

smelling route

route taken by air during quiet breathing

FIG. 24/2.—THE OLFACTORY NERVES AND CONNECTIONS

an outlying portion of the brain, is the slightly bulbous (enlarged) portion of the olfactory nerve tract which lies above the cribriform plate of the ethmoid bone. From the olfactory bulb sensation is passed along the *olfactory tract* by several relaying stations until it reaches the final receiving area in the olfactory centre which lies in the temporal lobe of the cerebral hemisphere where the sensation is interpreted.

The *sense of smell* is stimulated by gases inhaled or by small particles. It is a very delicate sense, and becomes easily deadened when expoced to any one odour for some time. Persons in a stuffy room rapidly become oblivious to the unpleasant odours, which strike others forcibly on entering the room from the fresh air. The sense of smell is also lessened if the nasal mucous membrane is very dry, very wet or swollen, as in a cold in the head. Smells are described as pleasant or unpleasant.

Entire loss of the sense of smell may complicate an injury to the head.

Short **Clinical Notes** will be found in the preceding pages.

Chapter 25

ORGANS OF SPECIAL SENSE—THE EYE AND SIGHT

The *optic* or second cranial nerve is the sensory nerve of sight. The nerve arises from the ganglion cells in the retina (*see also* Fig. 25/4) which converge to form the optic nerve. The nerve passes backwards and medially and runs through the optic canal (*see* Fig. 3/7, page 69), to enter the cranial cavity and thence to the optic chiasma. The optic nerve has three

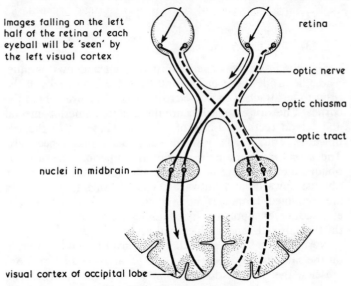

Images falling on the left half of the retina of each eyeball will be 'seen' by the left visual cortex

retina

optic nerve

optic chiasma

optic tract

nuclei in midbrain

visual cortex of occipital lobe

FIG. 25/1.—THE OPTIC TRACT

coverings, similar to the meninges of the brain. The outer one is tough and fibrous and blends with the sclera (*see* Fig. 25/2), the middle covering is delicate like the arachnoid mater and the inner one is vascular.

When the fibres reach the *optic chiasma* half of the fibres converge to reach the opposite side of the *optic tract* (*see* above).

By means of this arrangement of fibres each optic nerve is related to both sides of the brain. The *visual centre* lies in the cortex of the occipital lobe of the brain (*see* Fig. 22/7, page 335). The eyeball is the organ of sight. It is contained in the bony orbit (*see* page 71), and protected by appendages such as the eyelids, eyebrows, conjunctiva, and the lacrimal apparatus.

THE EYEBALL

The eyeball is generally described as a globe or sphere, but it is oval, not circular. It is about an inch in diameter, transparent in front, and composed of three layers:

(1) outer fibrous, the supporting layer
(2) middle, vascular, and
(3) inner nervous layer.

Six muscles move the eye, four straight and two oblique. These lie inside the orbit passing from the bony walls of the orbit to be attached to the sclerotic coat of the eye behind the cornea. The *straight muscles* are the superior, inferior, medial and lateral rectus muscles of the eye. These move the eye upwards, downwards, inwards, and outwards respectively. The *oblique muscles* are the inferior and superior. The superior oblique moves the eye downwards and outwards, the inferior oblique upwards and outwards. The movements of the eyes are combined, both eyes move to right or left, up and down, etc., the nerves supplying these muscles are the *motores oculi*—the third, fourth, and sixth cranial nerves.

Normally the axes of both eyes converge simultaneously on the same point; when, owing to paralysis of one or more muscles, they fail to do so *squint* or *strabismus* exists. The condition may be congenital or acquired. When it cannot be corrected by glasses and re-education, operation is undertaken followed by careful re-education and training.

The sclera is the tough outer fibrous coat. It forms the *white of the eye* and is continuous in front with a transparent window-membrane, the *cornea*. The sclera protects the delicate structures of the eye and helps to maintain the shape of the eyeball.

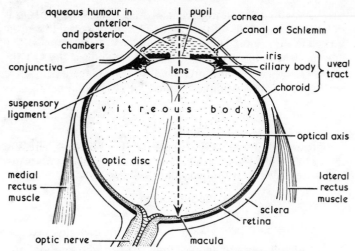

FIG. 25/2.—HORIZONTAL SECTION THROUGH THE RIGHT EYEBALL

The choroid or middle vascular coat contains the blood vessels, which are the ramifications of the ophthalmic artery, a branch of the internal carotid. This vascular coat forms the *iris* with the central opening or *pupil* of the eye. The pigmented layer behind the iris gives it colour and determines whether the eye is blue, brown, grey, etc. The *choroid* is continuous in front with the *iris* and just behind the iris this coat is thickened to form the *ciliary body*, thus the *ciliary body* lies between the choroid and the iris. It contains circular muscle fibres and radiating fibres; contraction of the former contracts the pupil of the eye.

Together these form the *uveal tract*, consisting of iris, ciliary body and choroid coat. Inflammation of the individual parts is described as *iritis*, *cyclitis* and *choroiditis* respectively, or collectively as *uveitis*. When one part of this tract is inflamed the congestion is quickly spread to an adjacent part.

The retina is the inner nervous coat of the eye, composed of a number of layers of fibres, nerve cells, rods and cones (*see* Fig. 25/4), all of which are included in the construction of the retina, the delicate nerve tissue conducting the nerve impulses from without inwards to the *optic disc*, the point where the optic nerve leaves the eyeball. This is a *blind spot*, as it possesses no retina. The most acutely sensitive part of the retina is the *macula*, which lies just external to the optic disc, and exactly opposite the centre of the pupil.

For retinal detachment *see* Clinical Notes.

In examining the eyeball from front to back the following parts are seen (please consult Fig. 25/2).

Cornea, the transparent front portion continuous with the dense white sclera. It consists of several layers. The superficial layer is stratified epithelium continuous with the conjunctiva.

Anterior chamber, between cornea and iris.

Iris, the coloured curtain in front of the lens which is continuous with the choroid coat. The iris contains two sets of involuntary or plain muscle fibres—one set contracts the size of the pupil and the other set dilates the pupil.

Pupil, the dark central spot which is an opening in the iris through which light reaches the retina.

Posterior chamber, between iris and lens. Both anterior and posterior chambers are filled with *aqueous humour*.

Aqueous humour of the eye. This fluid is derived from the ciliary body and it is re-absorbed into the blood stream at the angle between the iris and cornea by a tiny vein known as the canal of Schlemm.

Lens, a bi-convex transparent body made up of several layers. It lies just behind the iris. It has both in front and behind a membrane, known as the *suspensory ligament*, by which the lens is attached to the ciliary body. When the suspensory ligament is slackened the lens recoils and becomes thicker, when the ligament is taut the lens flattens. Slackening of the lens is controlled by contraction of the ciliary muscle.

Vitreous humour. The remaining back portion of the eyeball extending from the lens to the retina, is filled with a jelly-like albuminous fluid—the vitreous humour of the eye which serves to give it shape and firmness and to keep the retina in contact with the choroid and sclerotic coats.

Function of the Eye. The eye is the special organ of sight. It is constructed to receive the stimuli of rays of light on the retina, and, by means of the optic nerves, to transmit these to the visual centres of the brain for interpretation.

The cornea acts as a transparent window protecting the delicate structures behind it, and helping to focus images on to the retina. It does not contain any blood vessels.

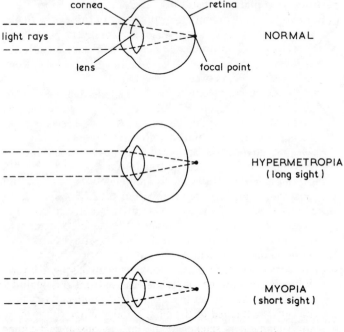

FIG. 25/3.—POINTS OF REFRACTION (EXAGGERATED)

The iris, with its central opening, the pupil, is a movable disc, which acts as a curtain to protect the retina, controlling the amount of light entering the eye.

The lens is the principal organ of focus, bending rays of light reflected from objects seen to a clear image on the retina. The lens is contained in an elastic capsule, attached to the ciliary body of the choroid by a suspensory ligament. By means of the ciliary muscle the anterior surface of the lens is made more or less convex, to focus near or distant objects. This is visual *accommodation*.

The pigmented choroid coat darkens the inner chamber of the eye, comparable to the blackened interior of a photographic camera.

The retina is the nervous mechanism of sight. It contains the endings of the optic nerves, and is comparable to a sensitive photographic plate.

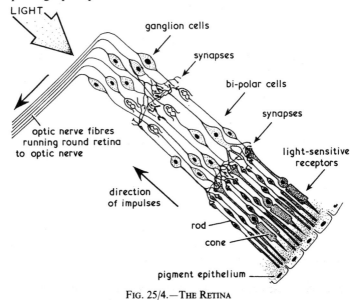

FIG. 25/4.—THE RETINA

A diagram showing the principal layers of the retina. Reading from below upwards will be seen the different structures by which sight is received and transmitted ·to the ganglion cells of which the axons form the fibres of the optic nerve.
(For description of nerve synapse *see* page 351.)

When an image is perceived, rays of light from the object seen pass through the cornea, aqueous humour, lens, and vitreous body to stimulate the nerve endings in the retina. The stimuli received by the retina pass along the optic tracts to the visual areas of the brain, to be interpreted. Both areas receive messages from both eyes, thus giving perspective and contour.

In an ordinary camera one lens is provided. In the eye, whilst the crystalline lens is very important in focusing the image on the retina, there are in all four structures acting as lenses: the cornea, the aqueous humour, the crystalline lens, and the vitreous body.

As in all interpretation of sensation from the surface, a number of relaying stations are concerned with the transmission of the sensation which in this case is sight. A number of these relaying stations are in the retina as can be seen by studying Fig. 25/4. Internal to the periphery of the retina are layers of rods and cones which are highly specialized sight cells sensitive to light. The circular interruptions in these are termed granules. The proximal ends of the rods and cones form the first synapse with a layer of bi-polar cells, still in the retina. The second processes of these cells form the second nerve synapse with large ganglion cells, also in the retina. The axons of these cells form the fibres of the optic nerve. These pass backwards, first reaching the lower centre in special bodies near the thalamus, and finally reaching the special visual centre in the occipital lobe of the cerebral hemisphere where sight is interpreted.

APPENDAGES OF THE EYE

Eyebrows. These are two arches of thick skin from which hairs grow. The eyebrows are attached to muscles beneath, and serve to protect the eye from too great light.

Eyelids. These are two plates, the *tarsal plates*, which are composed of very dense fibrous tissue, covered by skin and lined with conjunctiva. The tissue beneath the skin does not contain fat. The upper eyelid is larger than the lower, and is raised by the *levator palpebrae* muscle. The lids are closed by a circular muscle, the *orbicularis oculi*. Eyelashes are attached to the free margins of the lids, and protect the eyes from dust and light.

For inflammatory disorders *see* Clinical Notes.

Refracting function of the Eye. As already mentioned the rays of light falling upon the eye which will bring an image to focus on the retina pass through and are altered by the cornea, lens, aqueous and vitreous bodies, but the lens is the principal organ bending rays of light to focus an image on the retina. In the normal eye these rays converge to strike a point on the retina as illustrated in Fig. 25/3 where the image is focused.

Clinically *abnormalities of refraction* result in defects of visual accommodation either as the result of alteration in the shape of the eyeball or abnormalities of the lens. In *Hypermetropia* or long sight the eye is short from back to front and therefore the lens focuses the image *behind the retina*, whilst in *myopia* or short sight the eyeball is longer than normal from back to front and the lens focuses the image *in front of the retina*.

Astigmatism is an error of refraction which occurs when rays of light fall upon lines on the retina and not on sharp points. This is due to alteration in the curves of the lens and is cor-

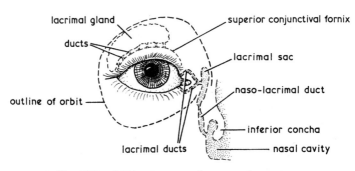

FIG. 25/5.—A DIAGRAM OF THE LACRIMAL APPARATUS

rected by spectacles with lenses which are convex in the direction which is lacking in the abnormal lens and so make good the deficiency.

Presbyopia is the term used to describe the defect of accommodation which occurs in advancing age. The lens loses its elasticity and becomes less resilient and fails to focus the image of a near object. Distant vision is unimpaired. The subject with presbyopia is seen to hold a paper a distance away in order to be able to read it. This defect is corrected by providing convex lenses.

Conjunctiva. This is the mucous membrane lining the eyelids and covering the front of the sclera. It is continuous with the lining membrane of the lacrimal ducts, the lacrimal sac and with the naso-lacrimal ducts. When the lids are closed the conjunctiva forms a closed sac. It is into this sac that *drops are instilled* into an eye. Drops should be placed at the outer side of the fornix which is the cul-de-sac where the conjunctiva which covers the eyeball is reflected upon the eyelid. The drops thus may exert their effect on the eye before being washed away in the tear ducts. The same precaution should be taken when *irrigating an eye*.

Lacrimal Apparatus. The lacrimal glands are compound race-mose glands, situated at the upper outer corner of the orbital cavity, and secrete the *tears*, which at the upper and outer margins of the eye are poured into the conjunctival sac from the ducts of the lacrimal glands. As the eyelids move in blinking, the tears are distributed across the surface of the eyeball. A considerable amount of this fluid is evaporated and any excess passes from the inner angle of the eye into the lacrimal ducts and then by the naso-lacrimal duct into the nose. The flow of tears is increased by irritants (tear gases, for example) and by emotion.

Clinically *lacrimal obstruction* may be due to slight congestion of the narrow tear ducts; when this tendency is experienced the condition is made worse on exposure to cold winds or to irritation. Treatment is difficult; sometimes dilating the duct by passing a fine probe through it may help, but when obstruction is persistent the lacrimal sac is excised.

Acute dacryocystitis is the result of infection of an obstructed lacrimal sac, and an abscess develops, characterized by painful red swelling below the inner canthus.

Clinical Notes

Enucleation of the eye may be undertaken after injury, because of a malignant growth, and sometimes when a blind eye becomes unbearably painful. This operation is a mutilating one and constitutes a deformity; therefore the most careful consideration is given as to its advisability, in consultation with the patient and his relatives. The precaution of marking the eye to be excised must be meticulously made; this precaution applies equally whenever amputation or excision of an organ on one side of the body is considered.

Eyelids may be the site of infection. A *stye or hordeolum* is infection of an eyelash follicle; the lash should be removed and the lesion treated by heat.

Meibomian cysts are sebaceous cysts on the margins of the lids which need evacuation and treatment.

Blepharitis is inflammation of the margins; the lids are red, sore and sticky. All crusts should be removed, followed by hot-spoon bathing before treatment is undertaken.

Eversion of the lids, or *ectropion*, may be the result of ulceration or of injury. Inversion, or *entropion*, results from contraction after ulceration or after injury; the eyelashes irritate the eye, causing pain. *Epiphora* is a flowing of the conjunctival fluid on to the cheek; it occurs in lacrimal obstruction (*see* p. 379), in eversion of the lids and in conjunctivitis. *Ptosis* is drooping of the upper eyelid.

Conjunctivitis, or inflammation of the conjunctiva, may be acute or chronic; it is due to a variety of organisms. The eye or eyes feel hot and gritty, the lids swell, the conjunctiva is red, the eyes water and there is intolerance of light, *photophobia*. Treatment is directed to the cause.

Trachoma, a form of conjunctivitis, is due to a virus infection of the conjunctiva, common in the Middle East. It is one of the major causes of blindness in the world.

Affections of the *Cornea*. Minor injuries generally heal if carefully attended; more serious ones may be the cause of *corneal ulceration*, which is extremely painful and needs treatment and protection. *Keratitis* is inflammation of the cornea.

Corneal grafting is employed to replace a scarred, opaque cornea by a normal one, obtained either from a recently enucleated eye, or from a recently deceased donor, thus supplying a clear cornea and making sight possible. Corneal banks are available.

The Lens. *Cataract*, which is an opacity of the lens, may be partial or complete. It may be congenital, due to injury or a complication of diabetes, but it is the *senile cataract* due to degenerative changes in ageing, which is most often seen. *Excision of the lens* by one of the available operations is performed. The preparation for this operation and the post-operative care are meticulously carried out.

Glaucoma is due to increase of tension in the eye, and may be acute or chronic. It is due to the fluid in the anterior chamber of the eye not being drained away and the tension thus produced causes pressure on the optic nerve, with gradually failing sight.

Acute glaucoma has a sudden onset, with acute pain which cannot go unheeded. It is treated by miotics to contract the pupil, applications of heat by spoon-bathing, and by the administration of diuretics to relieve intra-ocular tension. *Trephining* is performed by making a small perforation which enables the fluid in the anterior chamber to drain away permanently. It is an intra-ocular procedure and needs the same post-operative care as after the operation for cataract.

Simple glaucoma may unfortunately pass unnoticed for years and yet the characteristic intra-ocular pressure is gradually developing. The only treatment, once this condition is recognized, is the insertion of miotics for life which imposes a heavy discipline. It is essential that any patient presenting for a physical examination should have his eyes tested by an ophthalmic surgeon. Simple glaucoma is one of the commonest causes of *blindness* in this country.

Retinal detachment is serious, as the retina is the organ of sight. The condition is generally dealt with by surgery, which may be scleral resection; or by the application of diathermy to the sclera, thus making a minute burn, so that when adhesions form these will re-attach the retina to the sclera. Cryosurgery, the application of cold, has a similar effect.

The *post-operative period* in either case is longer than that for excision of lens, and the position in which the patient is to be nursed will be directed by the surgeon.

Blindness. There are over 10 million blind people in the world, and it is considered that more than half could have been saved if modern preventive measures and surgery had been available in time. In this country the greatest causes of blindness are accidents and simple glaucoma, mentioned above. Most of the blindness in the world occurs in tropical countries and includes trachoma, already mentioned, smallpox and onchocerciasis, a disease prevalent in Central Asia and Africa, carried by gnats near rivers and locally described as 'river blindness'.

One cannot discount the number of people injured, even to the extent of blindness, by having harmful acids thrown in their face by burglars and the like.

For short Clinical Notes on abnormalities of refraction, *see* page 378 and lacrimal obstruction and dacryocystitis, *see* page 379.

Chapter 26

ORGANS OF SPECIAL SENSE—THE EAR AND HEARING

The Ear is the organ of hearing. The nerve supplying this special sense is the eighth cranial or *auditory nerve*. The ear is divided into three parts, external, middle, and internal.

The External Ear consists of the auricle or *pinna*, which in some of the lower animals is large and movable, and helps to collect the sound waves; and the *external auditory meatus*, which leads in from the pinna and conveys the vibrations of sound to the tympanic membrane.

This canal is about 2·5 cm (1 inch) long; the outer third is cartilaginous and the inner two-thirds bony. The cartilaginous

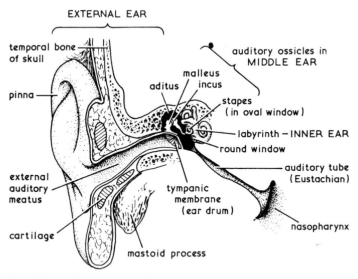

Fig. 26/1.—A Section of the Ear showing the parts which compose the External, Middle and Inner Ear

part is not straight and is directed upwards and backwards. It can be straightened by pulling the pinna upwards and back, which should be done when *syringing an ear*. The fluids should be directed towards the posterior and upper wall of the meatal canal. After syringing and inspection, any free fluid may be shaken out by the patient.

The auricle is irregularly shaped and is composed of cartilage and fibrous tissue except at its lowest point, the lobe of the ear, which is mainly fat. Three groups of muscles lie in front, above, and behind the ear, but only very slight movement of the auricle is possible in man.

The Middle Ear or *tympanic cavity* is a small chamber containing air, internal to the tympanic membrane or ear drum, which separates it from the external auditory meatus. It is narrow and has bony and membranous walls, and communicates behind by means of an opening called the *aditus* or attic with the mastoid antrum in the mastoid process of the temporal bone.

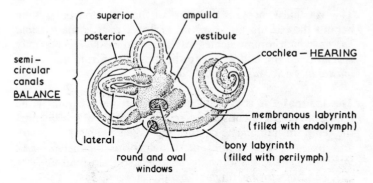

FIG. 26/2.—THE RIGHT LABYRINTH OF THE EAR

The Eustachian tube runs forward from the cavity of the middle ear into the naso-pharynx where it opens. Thus the air pressure may be equalized on each side of the drum, via the external auditory meatus, and via the Eustachian (pharyngo-tympanic) tube. The opening of the Eustachian tube is closed normally, but opens each time swallowing occurs. In this way the pressure of the air in the tympanic cavity is kept equal to that of the atmosphere and injuries and deafness due to greater or less pressure are avoided. It is this communication with the naso-pharynx that allows infection to spread from the nose or throat to the middle ear.

The auditory ossicles are three small bones arranged across the middle ear, like a chain reaching from the tympanic membrane to the inner ear. The external bone is the *malleus*, shaped like a hammer, the handle being attached to the tympanic membrane, while the head projects into the tympanic cavity.

The middle bone is the *incus* or anvil, which articulates with the malleus on the outer side, and with the innermost of the three ossicles, the stapes, on the inner side.

The *stapes* or stirrup bone is attached by its smaller end with the incus, and by its oval base with the membrane closing the *vestibuli fenestra*, or oval window. This chain of bone serves to transmit the vibrations of sound from the drum to the internal ear (*see* Fig. 26/1).

The Mastoid Process is that part of the temporal bone lying behind the ear; an air space in its upper part is the *mastoid antrum* which communicates with the middle ear. *Infection* may spread from the middle ear to the mastoid antrum and cause *mastoiditis*. (*See also* Clinical Notes.)

The Internal Ear is contained in the petrous portion of the temporal bone. It consists of several cavities which channel the temporal bone. The cavities are called the *bony labyrinth* and are lined with membrane, which forms the *membranous labyrinth*. These membranous channels contain fluid and the nerve endings for hearing and balance. (*See* opposite.)

The bony labyrinth consists of three parts.

The vestibule is the central part with which all the others communicate, as doors may open out of the vestibule of a dwelling.

The semicircular canals communicate with the vestibule. There are three of them, superior, posterior, and lateral; the latter is situated horizontally and all three of the canals lie at right angles to each other. Each canal has a swelling at one end called the *ampulla*. (It is by movements of the fluid stimulating special nerve endings within the ampullae that we are made conscious of our position. The function of this part of the internal ear is to assist the cerebellum in the control of the equilibrium, and the sense of the position of the body.)

The cochlea is a spiral tube twisted on itself resembling a snail-shell. The coils are arranged round a central bony cone-shaped axis called the *modiolus*.

Within each of these is a *membranous counterpart* to which are attached the endings of the auditory nerve. The fluid within the membranous labyrinth is called *endolymph*, the fluid outside it and within the bony labyrinth is the *perilymph*. There are two windows in this bony enclosure:

(1) the *fenestra vestibuli* (also called ovalis, as it is oval in shape) is closed by the stapes,

(2) the *fenestra cochleae* (also called rotunda, as it is round) is closed by a membrane (*see* Fig. 26/2).

Both these are directed towards the middle ear. The purpose of these windows in the bony labyrinth is to allow the vibrations transmitted from the middle ear to occur in the perilymph (fluid being practically incompressible). Vibrations of the perilymph are transmitted to the endolymph and so activate the nerve endings of the auditory nerve.

The Auditory Nerve is in two portions: one collecting sensations from the vestibular portion of the inner ear is concerned with equilibrium. The fibres from this pass to the vestibular nuclei at the junction of the pons and medulla and then on to the cerebellum. The cochlear portion of the auditory nerve is the *true nerve of hearing*. Its fibres are first relayed to a special nucleus immediately behind the thalamus and thence to the final receiving centre in the cerebral cortex which lies in the under portion of the temporal lobe of the cerebrum (*see* Fig. 22/7, page 335).

Lesions of the cochlear nerve result in nerve deafness; lesions of the vestibular nerve in vertigo, ataxia and nystagmus.

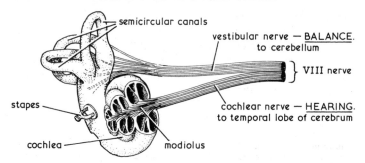

FIG. 26/3.—THE AUDITORY NERVE SHOWING THE VESTIBULAR AND COCHLEAR PORTIONS

Hearing. Sound is due to vibrations of atmosphere known as sound waves, varying in rate and volume. Sound waves pass along the external auditory canal and cause the tympanic membrane to vibrate. The malleus is attached to this membrane, and the vibration is transmitted through it to the incus and stapes. By movement on each other these bones magnify the vibrations, which are then communicated through the vestibular fenestra to the perilymph. Vibrations of the perilymph are transmitted through the membrane to the endolymph in the canal of the cochlea, *and the stimuli reach the nerve endings in the organ of Corti, to be conveyed to the brain by the auditory nerve.*

The sensation of hearing is interpreted by the brain as a pleasant or unpleasant sound, or one of noise or music. These terms are used in their widest sense. Irregular sound waves produce noise, regular rhythmic waves produce pleasant musical sounds. Sound travels at the rate of 343 metres (375 yards) per second in still air of 60°F (15–16°C). (For the effects of noise see opposite.)

Balance. *The vestibular nerve* distributed to the semicircular canals conveys to the brain the impulses generated there by alterations in the position of the fluid in these canals which have so much to do with the knowledge of the sense of the position of the head in relation to the body. If a person is suddenly thrown to one side, the tendency is for the head to bend towards the opposite side in order to maintain balance so that weight is adjusted, the erect position maintained, and a fall is prevented. It is the change in the position of the fluid in the semicircular canals which stimulates the impulse, obeyed as a reflex, by the quick response of the body to transfer weight and maintain equilibrium.

Clinical Notes

Infections of the Ear. The *External auditory meatus* may be the site of *furunculosis*, a boil or multiple boils in the canal, which can be extremely painful. Antibiotics are given and local treatment by gentle heat applied.

Blockage of the Eustachian tubes may be the result of infection or due to the presence of adenoids.

The Middle Ear. *Otitis media*, or infection of the middle ear, may follow influenza, measles and sinusitis. Antibiotics are given and heat applied locally.

Acute Mastoiditis may follow otitis media. The mastoid process is tender, painful and swollen, and there is a rise of temperature and increase in the pulse rate. It is comparatively rare but when it does arise, if the condition does not subside after antibiotic treatment, surgical intervention is undertaken.

Chronic suppurative otitis media and chronic mastoiditis may follow the acute infections.

Middle ear disease is fortunately rare at the present time in this country, but owing to its close proximity with the meninges, meningitis, extra-dural abscess, cerebral abscess and infection or thrombosis of the lateral sinus may arise as complications.

The Inner Ear. Two conditions may be mentioned: *Labyrinthitis*—usually due to spread of infection from the middle ear; in most instances the symptoms of giddiness, vomiting and deafness gradually pass off.

Ménière's Disease, is characterized by sudden attacks of giddiness, accompanied by deafness and tinnitus. In many cases the condition can be controlled by mild sedation and attention to the conditions of living.

Balance is sometimes temporarily disturbed after certain operations on the ear, such as stapedectomy, and in travel (motion) sickness; both are quickly corrected by one of the antihistamines contained in a number of proprietary preparations, such as Dramamine. Balance and gait (*see* page 386) may be permanently affected in an injury to the head.

The Effects of Noise are difficult to assess. Sleep is disturbed. In some the mental tension caused by noise results in increase in pulse rate and hypertension. Concentration and efficiency may be interfered with, leading to hazards; when a warning cannot be heard an accident may result.

The integrity of hearing may be affected by noise; intense noise leads to loss of hearing; exposure to continuous industrial and traffic noises results in loss of sensitivity of hearing, which may lead progressively to deafness.

Levels of noise expressed in decibels (db) compare the sound pressure levels, but do not register noise. Examples of these sound levels are: 60–70 decibels: ordinary conversation; 80–90: heavy traffic; 140–150: proximity of a jet engine.

The maximum level of noise the human ear can bear is 130 decibels, but people should not be exposed to this. Prolonged exposure to an intensity of 90 to 95 can injure hearing and those exposed to noise continuously, as in factories, should have regular hearing tests and be provided with ear protectors. Ear muffs and plastic ear plugs are available.

Deafness. The causes of deafness are too numerous to mention here, but the whole point of treatment is to direct it to the cause and to prevent the condition getting worse. The deaf are cut off from communication and may obtain relief from hearing aids or from surgery.

Deafness in an infant should be recognized as early as possible. Aids for the very young are available. A deaf child should be spoken to frequently, otherwise he may become mentally and socially retarded.

GLOSSARY OF EPONYMS

Many anatomical terms carry the name of the physician or surgeon who discovered them or whose work was specially connected with them. These eponyms are still sometimes used and are therefore listed here, although the modern practice is to use the anatomical name.

Alcock's canal Canal for the internal pudendal vessels in the ischiorectal fossa

Argyll-Robertson pupil A pupil that does not react to light but will react to accommodation

Auerbach's plexus Autonomic nerve plexus of the intestinal muscle

Bartholin's glands Small glands in the labia majora

Betz cells Giant pyramidal cells of the motor cortex of the brain

Bigelow's ligament The Y-shaped ileofemoral ligament of the hip-joint

Bowman's capsule Expanded proximal end of a renal tubule surrounding the glomerulus of the kidney

Bowman's membrane Anterior elastic membrane of the cornea

Broca's area Speech area of the brain

Brunner's glands Glands of the duodenum

Camper's fascia Superficial fatty layer of superficial fascia over the abdomen

Cloquet's node Lymph node in the femoral canal

Astley Cooper's ligament (1) Pectineal ligament; (2) suspensory ligaments of the breast

Corti, organ of Part of the internal ear

Cowper's glands Bulbo-urethral glands

Denonvilliers, fascia of The prosto-peritoneal aponeurosis

Descemet's membrane Posterior elastic membrane of the cornea

Douglas, pouch of Recto-vaginal peritoneal pouch

Dupuytren's contracture Contracture of the palmar fascia

Edinger-Westphal nucleus The parasympathetic nucleus of the third cranial nerve

Eustachian tube Pharyngo-tympanic tube of the ear

Fallopian tubes The uterine tubes

Gimbernat's ligament The lacunar ligament of the inguinal ligament

Graafian follicles Ovarian follicles

Haversian canals Canals containing blood within compact bone

Henle, loop of The looped part of the uriniferous tubule

His, bundle of Atrioventricular bundle

Houston, valves of Valve-like structures in the anus

Hunter's canal Subsartorial canal

Jacobson's nerve A tympanic branch of the glossopharyngeal nerve

Kupffer cells Stellate cells lining the sinusoids of the liver

Langer's lines Cleavage lines of the skin

Langerhans, islets of Clumps of cells lying in the interalveolar tissue of the pancreas

Lieberkühn, crypts of Intestinal glands

Lister's tubercle Prominence on the posterior surface at the lower end of the radius beside the groove for the tendon of extensor pollicis longus

Louis, angle of Angle formed by the junction of manubrium and body of the sternum

Luschka, foramen of Foramen in the lateral recess of the fourth ventricle

Mackenrodt, ligament of Transversus colli ligament of uterus

Malpighian corpuscles Splenic corpuscles

Meckel's cartilage Embryonic cartilage of the first branchial arch

Meckel's cave Recess for the trigeminal ganglion

Meckel's diverticulum Remains of the omphalomesenteric duct

Meibomian glands Sebaceous follicles of the eyelids

Meissner's plexus Submucous nerve plexus of the intestine

Monro, foramen of Interventricular foramen between the lateral and the third ventricle of the brain

Morgagni, columns of Mucosal folds of the anal canal

Mullerian duct Primordial female genital tract

Oddi, sphincter of Sphincter at lower end of the bile duct

Pacinian corpuscles End-organs of sensory nerves

Peyer's patches Aggregated lymph follicles in the ileum

Poupart's ligament The inguinal ligament

Ranvier's nodes Gaps in the medullary sheath of myelinated nerves

Rathke's pouch A depression in the roof of the embryonic mouth in front of the buccopharyngeal membrane

Retzius, cave of The pubovesical space

Rolandic fissure Central sulcus of the cerebral cortex

Rutherford Morrison's pouch The hepatorenal pouch of peritoneum

Scarpa's fascia Fibrous deep layer of superficial fascia over the abdomen

Scarpa's triangle Femoral triangle

Schlemm, canal of Canal at the junction of cornea and sclera

Schwann cells Cells of the neurolemma

Sharpey's fibres Connective tissue fibres between the periosteum and bone

Shrapnell's membrane Flaccid portion of tympanic membrane

Sibson's fascia Membrane covering the apex of the lungs

Stensen's duct Duct of the parotid gland

Sylvian fissure Lateral cerebral fissure

Sylvius, aqueduct of Canal joining the third to the fourth ventricle within the brain

Tenon's capsule Fascia bulbi of the eye

Treves, bloodless fold of Bloodless fold of peritoneum which runs from the terminal ileum towards the base of the appendix and becomes attached to the meso-appendix

Vater, ampulla of Ampulla of the common bile duct

Vidian nerve Nerve of the pterygoid canal

Waldeyer's fascia Loose areolar tissue surrounding the rectal venous plexus

Waldeyer's ring Ring of adenoidal tissue of the nasopharynx

Wharton's duct Duct of the submandibular salivary gland

Wharton's jelly Embryonic connective tissue of the umbilical cord

Willis, circle of Arterial circle at the base of the brain

Winslow, foramen of The foramen of entry into the lesser sac of peritoneum

Wolffian duct The primordial ureter

INDEX

393

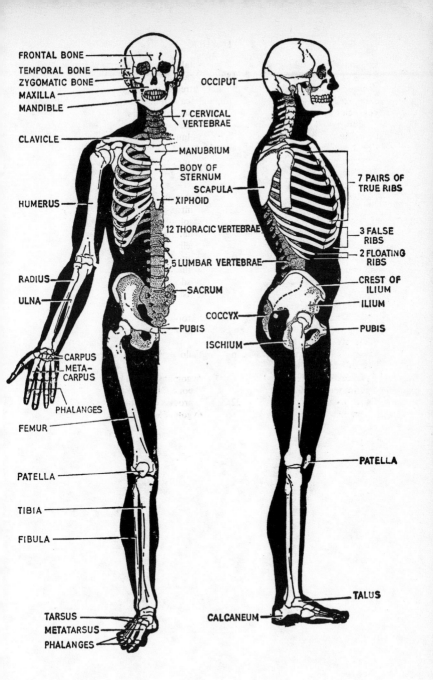

FRONTAL BONE
TEMPORAL BONE
ZYGOMATIC BONE
MAXILLA
MANDIBLE

OCCIPUT

7 CERVICAL
VERTEBRAE

CLAVICLE

MANUBRIUM

BODY OF
STERNUM

SCAPULA

XIPHOID

7 PAIRS OF
TRUE RIBS

HUMERUS

12 THORACIC VERTEBRAE

3 FALSE
RIBS

2 FLOATING
RIBS

5 LUMBAR VERTEBRAE

RADIUS

ULNA

SACRUM

CREST OF
ILIUM

COCCYX

ILIUM

PUBIS

PUBIS

ISCHIUM

CARPUS
META-
CARPUS

PHALANGES

FEMUR

PATELLA

PATELLA

TIBIA

FIBULA

TALUS

TARSUS
METATARSUS
PHALANGES

CALCANEUM

NOTES

NOTES

NOTES

NOTES

NOTES